NARRATIVES AND SPACES

NARRATIVES AND SPACES

Technology and the Construction of American Culture

David E. Nye

Columbia University Press, New York

Columbia University Press
Publishers Since 1893
New York   Chichester, West Sussex
Copyright © 1997 Columbia University Press

First published in 1997 by University of Exeter Press in Great Britain,
in the series *Representing American Culture* edited by Mick Gidley.

Library of Congress Cataloging-in-Publication Data

Nye. David E., 1946–
    Narratives and Spaces: technology and the construction of
American culture/David E. Nye.
        p.   cm.
    Includes bibliographical references (p.191) and index.
    ISBN 0–231–11196–7 (cloth). — ISBN 0–231–11197–5 (pbk.)
    1. United States—Civilization—19th century. 2. United
States—Civilization—20th century. 3. Technology and civilization.
4. Technology in literature. 5. Technology—Social aspects—United
States—History.   6. Public spaces—United States—History—19th
century.   7. Public spaces—United States—History—20th century.
I. Title.
E169. 1.N816   1998   97-28777
973—dc21   ∞

Casebound editions of Columbia University Press books are printed on
permanent and durable acid-free paper

Printed in Great Britain by Short Run Press Ltd, Exeter

c 10 9 8 7 6 5 4 3 2 1
p 10 9 8 7 6 5 4 3 2 1

For my uncle,
George Nye,
who encouraged me when I was young.

Contents

Illustrations

Foreword by Mick Gidley

The poet and scholar Charles Olson opened *Call Me Ishmael* (1947), his perceptive study of Herman Melville, with the famous words "I take SPACE to be the central fact to man born in America, from Folsom cave to now. I spell it large because it comes large here. Large and without mercy". David Nye's *Narratives and Spaces: Technology and the Construction of American Culture* looks at space not only as a "fact", but also as it is figured—and as it itself has helped to forge—aspects of American culture in the twentieth century. Technology—the other key word in this book— has also had a powerful formative role in American culture. This is so whether we see it in linear chronological terms akin to Olson's, from the Native American-fashioned projectile points found in pre-Ice Age fossilised mammoths to present-day space projects, or in symbolic terms, such as Henry Adams' oft-cited vision of "the Virgin", with all her attendant values, set against the ultimate triumph of "the Dynamo" he saw at a world's fair, a juxtaposition elaborated in *The Education of Henry Adams* (1907) and played upon by others in numerous later formulations, such as the seemingly oxymoronic notion of the skyscraper as a "cathedral of commerce". In his book, just as he did in his deservedly much-praised *Electrifying America* (1990) and *American Technological Sublime* (1994), David Nye examines examples of machine-based change and probes both the emergence of new conceptual categories to which such changes gave rise and the stories—the narratives, in every sense—by which they were given currency and credence in the culture.

Acknowledgements

I want to thank Mick Gidley who not only suggested this project but also served as editor. Many others were helpful, and at the risk of leaving out some who deserve mention, I want first to thank those whose invitations to lecture quite literally made many of these chapters possible: Tony Badger, Rob Kroes, Gerhard Hoffmann, Alfred Hornung, Paul Levine, and the European American Studies Association. Many colleagues offered useful suggestions and references, including Robert Rydell, Dale Carter, Carl Pedersen, Ronald Johnson, Terry Sharrer, and Svend Erik Larsen. Stuart Kidd and Mick Gidley were both extremely conscienscious and thoughtful readers of the entire manuscript, and their proposed revisions greatly strengthened the result.

My research was conducted in many libraries and collections over the past decade. I thank the following: Library of Congress, Washington, DC; National Aeronautics and Space Administration Library, Washington, DC; Warshaw Collection, National Museum of American History, Washington, DC; National Archives, Washington, DC; Edison Pioneer Papers, Henry Ford Museum Library, Dearborn; Duke University Library, Special Collections, J. Walter Thompson Corporate Archives; National Rural Electric Cooperative Association Library, Washington, DC; General Electric Library, Schenectady; General Electric Photographic Archives, Schenectady, New York; Omaha Public Library; Rochester Public Library; New York Public Library, World's Fair Collection; John F. Kennedy Library, Berlin; Hagley Museum and Library, Delaware; Visitor Center Library, Grand Canyon National Park; the British Museum Library; and the Odense University Library, Denmark.

This book was researched primarily in the United States, but written in Europe. While none of the chapters has appeared in the United States, one appeared in England, four in Denmark, three in The Netherlands and one in Germany. One was given as a conference paper at Sidney Sussex College, Cambridge. All have been revised for this volume.

Chapter Attribution

Chapter 1 appeared in *American Images* (Copenhagen: USIS, 1995), pp. 183–204.

Chapter 2 in a shorter version appeared in Rob Kroes, ed., *The American West, As Seen by Europeans and American* (Amsterdam: Free University Press, 1989), pp. 183–202.

Chapter 3 appeared in a special issue of *Word & Image* in 1988.

Chapter 4 is based on an unpublished conference paper, delivered at Sidney Sussex College, Cambridge University, September, 1994.

Chapter 5 appeared in considerably different form in *American Studies in Scandinavia*, 25:2, 1993, pp. 73–91.

Chapter 6 in somewhat different form came out as "Ventriloquist for the Past, E. L. Doctorow's World's Fair," Prepublication of the Odense University English Department, No. 48, 1989, pp. 1–26.

Chapter 7 appeared in a shorter version in Rob Rydell and Nancy Gwinn, *Fair Representations: World's Fairs and the Modern World* (Amsterdam: Free University Press, 1994), pp. 140–156.

Chapter 8 was in Rob Kroes, et al., *Cultural Transmissions and Receptions: American Mass Culture in Europe* (Amsterdam: Free University Press, 1993), pp. 47–64.

Chapter 9 was issued as a prepublication, by the Odense University Center for American Studies, as OASIS, No. 19, 1995, pp. 1–18.

Chapter 10 first appeared in Gerhard Hoffmann and Alfred Hornung, eds. *Affirmation and Negation in Contemporary American Culture* (Heidelberg: Universitetsverlag C. Winter, 1994), pp. 313–328.

Introduction

Technology is generally understood to be a central part of modern life, especially in that most technological of cultures, the United States, and yet how this came to be so has only recently become a historical subject. The first journal in this field, *Technology and Culture*, began as recently as 1959, more than a decade after the atomic bomb. Graduate education in the area did not start until after the Apollo XI landing on the moon. Yet if the research subject emerged slowly, public fascination with technology has been widespread since the middle of the nineteenth century, as the following chapters amply demonstrate. Between 1830 and 1850 literally millions of Americans rode the new railroads for recreation; millions flocked to world's fairs to see the latest machines; and in the past two decades millions more rushed to buy personal computers as they came on the market. This book examines these and other examples of technology in society, including the white settlement of the western United States, rural electrification during the New Deal and after, the 1970s energy crisis, and the public response to the Apollo moon landings.

Through these examples, I seek to discredit the idea of technological determinism,[1] and to argue that machines are social constructions which Americans long have built into both their narratives and their sense of place. Technologies are central parts of American self-representation, tourism, narrative practice, and visual sensibility. Americans have appropriated and developed machines in their own way, and woven them into landscapes, social relations, and a sense of history. Technologies of transportation and representation were inseparable from the construction of "natural" tourist sites at Niagara Falls and the Grand Canyon, and likewise, the American sense of urban space is inseparable from the automobile, the freeway, and the skyscraper. People use technologies to reshape and reimagine their material context, and their experience of any space is a complex, mediated encounter. Machines do not simply appear. People invent and shape them within a larger context, which includes visual practices and narrative strategies.

As Leo Marx has recently pointed out, the term "technology" has only been widely used in the twentieth century.[2] While suggested by a Harvard professor, Jacob Bigelow, as early as 1828 as an all-encompassing word for systems of mechanical improvements, most Americans long continued to speak of "the mechanic arts" or simply "the arts". The new term only gradually became common, and was first used widely at the end of the nineteenth century, during the rapid expansion of engineering education. It had a positive connotation in writers such as Thorstein Veblen and other advocates of a larger role for engineers in American society. In more recent use, however, the term tends to obscure human action and to represent machines as an abstract force in history. We often read in the newspapers that "technology" causes change. One of the emphasis in this volume will be that machines are not autonomous. Particular corporations and individuals were responsible for irrigating the West, lighting the world's fairs, installing computers in factories, or mining coal, iron ore, and other resources. Yet if human beings decide which machines to use and how to use them, this does not mean that people collectively know what they are doing. Quite the contrary, the combined choices of inventors, entrepreneurs, workers, and consumers create unanticipated cultural consequences. Because no single individual or institution wills certain changes, they are often mistaken for attributes of the machine itself. And once machines are perceived as active forces in their own right, they become part of narrative.

Too often writers minimize the complexity of technological change, either by assuming that corporations exercise hegemonic control, or by assuming that machines inexorably shape historical events. Even if some people undeniably have more power than others, those who initially control new technologies often find it difficult to maintain their dominance. The railroad once was the hegemonic American form of transportation; IBM once appeared invincible in the world of computers. But such apparent invincibility did not last. Even the enormous prestige of the space program during the Cold War, abetted by saturation media coverage, did not hold public interest for long. Machines are human-directed agents of cultural change. Human beings invent, market, and use technologies to create new sensibilities, and consumers in turn make demands on them.

Technologies are contested terrains. World's fairs, spectacular lighting, and media events all have changed in response to shifting public taste. Americans had divided reactions to the Apollo Space Program, and they created divergent narratives to understand computers, the electrification of society, and the meanings of energy shortages. Technologies are contested, and a close reading of such events as a world's fair or a space launch reveals not hegemonic control but a welter of meanings. The public understands technologies not as one-dimensional agents of change, but as both enabling and disabling innovations. Photography, railroads, rockets, electricity, and computers all multiply the possibilities of perception and of cultural experience, yet each also unbalances social relations. Overall, I take technology to be part of an ongoing conversation between generations and between social groups over their differing conceptions of what is desirable, possible or even real. Emil Durkheim understood that, "when ... conflicts break out ... they do not take place between the ideal and the real but between different ideals, between the ideal of yesterday and the ideal of today."[3]

Technologies are part of a dialogue between human beings about their differing perceptions. This dialogue takes the form of narratives, different stories we tell each other to make sense of the transformations that accompany the adoption of new machines. These narratives may focus on an older world that is fading into the past, as in the case of Wright Morris' meditations on the rural Midwest. Or they may engage the future, projecting utopian visions of ease and abundance, as was the case with the New York World's Fair of 1939. But whatever the narrative form, machines are seldom understood by the public as purely abstract things-in-themselves. Rather, technologies function as central parts of dramatic events.

This dramatization often takes place in a rather abstract American sense of space, which is conceived of as a *tabula rasa*. Thus the American West was long perceived as "virgin land," an empty region waiting to be appropriated.[4] This peculiar sense of space is legible in the boundaries of the western states. Mountainous Wyoming and Colorado are virtually square; their boundaries have nothing to do with local topography and everything to do with geometry and the compass. Likewise, a world's fairground is typically conceived of as an empty space, walled off from the rest of society.[5] Within this compound an ideal world can be erected, pointing the way to a more perfect future. The same impulse was manifested in the myriad new towns laid out in the developing nation during the nineteenth century. Through forested tracts and unpeopled swamps surveyors would mark off streets and lots, to be sold off with the expectation that an important metropolis would rise on the site. One traveler through the rural South in 1828 described a scene repeated in all parts of the nation.

> The first thing to which our attention was called was a long line cut through the coppice-wood of oaks. This our guide begged us to observe was to be the principal street; and the brushwood having been cut away so as to leave a lane four feet wide, with small stakes driven in at intervals, we could walk along it easily enough. On reaching the middle point, our friend, looking around him, exclaimed in raptures at the prospect of the future greatness of Columbus: 'Here you are in the center of the city!' He assured us further, that, within a very short period, this pathway would be converted into a street sixty yards wide ... After threading our way for some time amongst the trees, we came in sight, here and there, of huts made partly of planks, partly of bark....[6]

Many such plans came to nothing, but five years later this particular town had more than 2,000 inhabitants and was growing rapidly. A few years later Michel Chevalier was astonished to find that the land around "a little town by the name of Chicago" had been laid out as a vast city, and that despite a population of only 2,000, "the land for twenty-five miles around has been sold, resold and sold again in small sections, not, however, at Chicago, but at New York." In an as "yet uncultivated wilderness" Americans were investing in lots sufficient to house 300,000 people, "more than any city of the New World at present contains."[7] Two generations later Theodore Dreiser described in *Sister Carrie* the continued expansion of Chicago into the prairie after its population had surpassed half a million.[8] Today the driver approaching Las Vegas finds rock-strewn desert lots being sold as prime suburban real estate. It would seem that in the American imagination first there is an empty space traversed by a grid of surveyor's lines, followed by the dramatic imposition of human will on this space. An act of imagination and expropriation creates a landscape. And if the creation does not suit, it can be erased and something new erected on the spot, or it can be abandoned to grow up with new trees.

Technologies are always involved in the creation of a landscape. On one level, this is obvious enough, as forestry, agriculture, road-building, or river navigation involve some transformation of physical space to serve human requirements. All landscapes are partially shaped by people, and

Figure 1: San Diego Exposition, 1915
(General Electric Photographic Archives)

while it might seem that this generalization is far less true in such places as Utah, Nevada, and Arizona than in Connecticut or Florida, even these landscapes are socially constructed in several senses. First, as J. B. Jackson emphasizes, people always have left their mark on the land, both intentionally and inadvertently. Native Americans hunted and farmed North America for millennia. Since then white Americans have constructed mines, roads, farms, towns, dams, and irrigation canals that have changed the appearance of the land. Second, landscape views are socially constructed. The visual conventions we use to understand it emerged from painting and photography and were made universal by advertising and film. The tourist arrives with preconceptions. There is no innocent eye. Even if a site be regarded primarily as an unspoiled natural view, it is always seen through the lens of a powerful visual culture.

Furthermore, as Jackson argues, landscape refers to more than a view. The word "landscape" is also a verb, referring to the active process of changing the appearance of the world, and this landscaping occurs incessantly, though usually not as a meticulously planned activity. The result is a concatenation of many choices, ranging from state laws to the decorative whims of property owners who introduce new plants into a region. This volume will rely on Jackson's definition of landscape: "A composition of man-made or man-modified spaces to serve as infrastructure or background for our collective existence."[9] Landscape is not merely something seen, rather it is part of the essential infrastructure of existence. Landscape is a shared creation and a collective responsibility. Such a definition lends itself not only to studies of the countryside but also to studies of urban space.[10]

As Americans use machines to transform space into landscape (or to reconfigure an older landscape), they also construct narratives to make sense of this activity. These narratives cannot be taken at face value, and historians cannot be content with offering a mere synthesis of stories into one master narrative. The chapters that follow engage the problematics of writing when all forms of representation are under attack. This problem continually reappears, whether discussing the construction of the tourist gaze at the Grand Canyon, photography in the fiction of Wright Morris, deployment of energy narratives, the peculiar blend of history and fiction in the work of E. L. Doctorow, multiple narratives about specific New Deal programs, European self-representations at a world's fair, or the textualization of factory production facilitated by the computer. As this list immediately suggests, I prefer to put theory into a dialogue with particular instances. Each chapter begins within a certain historical moment, and then tests the claims of theory against the certainties, such as they are, of particular localities and moments. While I have tried to avoid the jargon of recent criticism, this volume does take seriously Jacques Derrida's deconstruction, critical theory, Fredric Jameson's reconceptualization of Marxism, and the meta-historical writings of Hayden White. Reading these authors not only raises questions about representation and narration, but also suggests new areas of inquiry, such as the electric landscape, energy narratives, and the representation of history at world's fairs. These essays accept their challenges to conventional historical writing. If determinism is dead, so too is history written like a realist novel of a century ago.

Yet if one important task for the present generation of historians is to dismantle the discursive apparatus of realism, writing history presents different problems than writing literary criticism. This difference arises from the materials engaged. Literary texts arrive on the desk as highly contrived works, which usually have the appearance of being complete. The critic can then dissect this corpus using an array of theories. Historians may agree that realism is an artefact of late nineteenth-century consciousness. They may reject the possibility of realistic representation.

Yet they inevitably begin not with a completed literary work, or a neatly delimited corpus of texts, but rather with extremely dispersed and discontinuous documents, located in different archives and libraries, to which must be added surviving elements of material culture. To deconstruct these scattered sources would atomize them into a vast field of fragments.

History is not virtually the same as fiction. Some things could not have happened in the past, though they may appear in novels. To give four examples: George Washington never drove an automobile; Theodore Roosevelt was not president after Woodrow Wilson; life expectancy of African-Americans under slavery was lower than at present; Alexander Graham Bell did not invent television. It is possible to establish which machines existed at a certain time, who governed when, how long people usually lived, or who did (or did not) invent something. These examples may seem trivial, but they are not. They set limits within which historical arguments can be developed. I cannot plausibly argue that Thomas Edison was a Frenchman or that sewing machines were unknown to Americans in 1885. When writing about the past there are many uncertainties, but this does not mean that one can say anything one pleases.

Yet history which focuses exclusively on the factual level cannot tell much of a story; it can only establish a sequence. All narratives must do their best to respect this sequence in order to remain plausible. The facts alone are potentially limitless and could include everything known about the past, without making distinctions between the items on the list. Once the historian selects texts, objects, photographs, films, and other documents to form a series and links these in an argument, he or she has begun to create a narrative.

Hayden White has argued that rewriting history does not make the past more solid: "Each new historical work only adds to the number of possible texts that have to be interpreted if a full and accurate picture of a given historical milieu is to be faithfully drawn. The relationship between the past to be analyzed and historical works produced by analysis of the documents is paradoxical; the more we know about the past, the more difficult it is to generalize about it."[11] Yet it does not follow that historical arguments should be regarded as rhetorical exercises. To say that the past is not perfectly knowable does not make it infinitely malleable.

Historians can learn a good deal from recent theory, but they are likely to agree with David Harvey's assessment.

> In their suspicion of any narrative that aspires to coherence, and in their rush to deconstruct anything that even looks like a meta-theory, they

[deconstructionists] challenged all basic propositions. To the degree that all the narrative accounts on offer contained hidden presuppositions and simplifications, they deserved critical scrutiny, if only to emerge the stronger for it. But in challenging all consensual standards of truth and justice, of ethics, and of meaning, and in pursing the dissolution of all narratives and meta-theories into a diffuse universe of language games, deconstructionism ended up, in spite of the best intentions of its more radical practitioners, by reducing knowledge and meaning to a rubble of signifiers.[12]

If historians cannot pretend to a complete objectivity, they can establish a bedrock of facts that set limits to what is true and what is false. Matters of chronology, measurement, kinship, and census-taking may be open to debate, but usually they can be established. Whatever the problems of representing such facts in language as part of an argument, historians work not with a "rubble of signifiers" but a brickyard of facts.

The selection and transformation of these facts into a narrative is a subjective process, however. For example, in chapter two I argue for the primary importance of technological change to explain the development of the American West. To arrive at this conclusion, my procedure was that which White described in *Metahistory*.[13] I began by prefiguring the ground of my investigation, in an act of imagination that was essentially poetic, as he says it must be. Yet this being admitted, as a historian I nevertheless must show that a wide and convincing range of facts and documents supports my prefiguration. It did not arise from sheer speculation, but from years of reading, teaching, traveling, and thinking. I ruled out many conceivable narratives because they were inconsistent with my accumulating knowledge. Prefiguration of a narrative is not the same as novelistic invention. Ultimately, a history will only be accepted as convincing if there is a good fit between the poetic act and the stubborn materials in the historian's brickyard.

A thorough deconstructionist would likely argue that historical documents cannot refer to the world, but only to other texts. Here we part company. Social historians do not need the wrecking ball of deconstruction. They labor to create new texts out of fragments, and in that sense they have been postmodernists for a long time. They usually foreground the artificiality of their construction. History offers not the absorption of fabulation, but rather immersion in an intertextual enterprise, whose margins are demarcated through footnotes, and whose reference points are often physical locations. Granting that historians have biases and ideologies, that we are faced with problems of representation, and that we necessarily prefigure the ground of our investigations, nevertheless, we are not writing fiction. While we may seem to be caught in a hall of mirrors, in which every writing is only a rewriting or a transmutation of previous codes, we might better be seen as architects endowed with a large stock of reliable materials and narrative designs.

What then is the design of this present volume? Clearly it is not a conventional narrative focused on a single set of events unfolding in linear time. Rather, it is organized thematically, and divided into three parts: spaces, narratives, and narratives in space. It begins by considering the social construction of two of the most famous "natural sites" in the United States, Niagara Falls and Grand Canyon, and how two distinct forms of the tourist gaze emerged from the different technologies of transportation and representation of the nineteenth and twentieth centuries. However, as the second chapter makes clear, many Americans were loath to understand their movement into the West as the technological conquest of space, and preferred to believe in Turner's frontier thesis. My purpose here is to underline the centrality of the railroad and electrification in the white settlement of the West and the production of its landscapes. The final example in this section considers a single case in detail. By 1940 the family farms and small towns of the Great Plains had come to be regarded as impoverished backwaters. Wright Morris understood one Nebraska area differently, as an outpost of rural values and aesthetic harmonies that were rapidly becoming inaccessible to urban Americans. Morris confronted this imminent erasure of cultural memory in *The Home Place*, a unique combination of prose and uncaptioned photographs that evokes the details of a landscape that farmers had made out of the abstractions of surveyors, railroad lines, and standardized goods.

After treating this book, which is thoroughly rooted in place and its daily routines, part two turns to narrative. The first section takes up the ways in which Americans made sense of rural decline in the 1930s and the rebirth promised by the introduction of electricity. If the initial conquest of the land had required railroads, windmills, barbed wire, steel tools, and new agricultural machinery, after 1935 electrification made possible a new countryside with urban amenities. Americans explained these changes, both during the New Deal and after, through four different narratives. The next chapter then considers the question of narrative in a more theoretical framework, examining different narratives about the cultural meaning of energy. Finally, this section ends, like part one, with a detailed examination of a single literary example, E. L. Doctorow's *World's Fair*.

The purpose is to confront head-on the relationship between history and fiction. Throughout this section Hayden White's *Metahistory* serves as the theoretical point of departure.

Part three examines four case studies of the intersection of narrative and space. Two are about failure: one the failure of European governments to construct world's fair exhibits that appealed to the American technological sense of space; the other, the failure of NASA to construct a convincing narrative to sell its space exploration program. The other two examples are about the successful integration of new machines into the American sense of space: a short history of the lighting of world's fairs, and a consideration of how Americans came to embrace the computer, and with it a new form of abstraction, cyberspace.

Odense, March, 1997

I

SPACES

ONE
Constructing Nature: Niagara Falls and the Grand Canyon

As airline and tourist brochures continually remind us, Niagara Falls and the Grand Canyon remain central icons of the United States. It was not always so. At the time of the American Revolution Niagara was on the frontier, too far away for most to visit. It became a national icon in the new nation one generation later. The Grand Canyon has an even shorter history as an icon. It was unknown to most citizens until the end of the nineteenth century. How did these places become tourist sites? The experience of "nature" produced at a given time cannot be understood as an isolated transaction between a tourist and an object of contemplation, for tourism expresses in its choice and framing of objects a particular historical moment. To make these observations more concrete, consider two such moments: Niagara Falls in the middle of the nineteenth century and the Grand Canyon in the late twentieth century. No attempt will be made to describe either of these sites, for this essay is less about them than about the self-conscious creation of points of view.[1]

I

Henry James commented in 1888: "When Americans went abroad in 1820 there was something romantic, almost heroic in it, as compared with the perpetual ferryings of the present hour, the hour at which photography and other conveniences have annihilated surprise."[2] James himself could not recall 1820, but rather imagined its circumstances. Photography was not invented until 1839, and its replication of the world in images was a novelty for the following generation. Before then, a traveler arriving in a new place had at best seen an engraving or a historical painting of the tourist object; more often verbal descriptions sufficed. Many sites suddenly appeared to an unprepared traveler, and had considerable force precisely because they were not anticipated. By the time that James traveled abroad, however, photographs had "annihilated surprise." Just as important were the "other conveniences" which rendered travel easier. James lived in a world of steamships, railways, comfortable carriages, porters, and international hotels. The tourist was swathed in a thick layer of Victorian comforts and moved at a stately pace through an itinerary of sights so well established that it was open to parody as early as 1869, when Mark Twain published *The Innocents Abroad*.[3] The "perpetual ferryings" James complained of in 1888 were the privilege of an extremely small upper class who could afford to take months to tour, with at least two weeks consumed by the transatlantic round trip alone. (To put this in perspective, even in 1997 the average American has only two weeks of vacation a year.[4]) In leisurely circumstances the nineteenth-century traveler had ample opportunity to prepare an itinerary, read about the places to be visited, and become acquainted with fellow tourists. The high cost of such a journey prohibited virtually all Americans from embarking on it, but even the small group that could go was organized into a hierarchy based on social standing and international connections. The purposes of such a trip, which cannot be separated from seeing the sights, were to acquire culture, round out one's education, and achieve social distinction.

Within the United States tourism had a less exclusive character for several reasons. First and most obviously, Americans could do a good deal of sight-seeing as a by-product of moving about the country. By making a detour or a weekend stop, many could manage to see Niagara Falls, the Natural Bridge of Virginia, Mamouth Cave, or the White Mountains.[5] Second, Americans had access to the world's most extensive system of railroads, which was built rapidly between 1830 and 1870, culminating in the completion of the first transcontinental line in 1869.[6] This development spread from the east coast inland, and by 1840 the United States had twice as much track as all of Europe.[7] After 1840 competition between midwestern cities ensured rapid expansion of the railways across the great plains.

Promoters early recognized that sight-seeing traffic was essential to turning a tidy profit. American railways were privately owned and often duplicated services. Several lines connected Chicago with New York, or Pittsburgh and St. Louis, for example. As a result, American fares were low and encouraged tourism. To promote business, railway executives issued calendars with views of the cities and scenery along their lines. They hired painters and photographers to illustrate them, and prepared booklets suggesting possible tourist routes.[8] Active promotion spread tourism to new classes of people, to those who had never been such extensive travelers in other nations.[9]

Railway travel literally changed the way that people saw the landscape. The passenger cars were slightly elevated above the surrounding country-side, and this fact, combined with the speed of the train, made it difficult to look at anything nearby. John Stilgoe notes that at "thirty miles an hour, everything within the thirty or forty-foot mark appears blurred, unless the traveler is willing to swivel his head as the train passes. Increases in speed force the observer to look ever further from the car, and particularly east of the Mississippi River, such long views are rare."[10] The railway journey erased the foreground and the local disappeared from the traveler's experience, while only a few scenes appeared worthy of notice. This editing of the landscape, framed by the windows of railway cars, transformed the journey into the opportunity to see a limited number of sites. These were memorialized in guide books, continually re-photographed, and, toward the end of the century, reproduced as postcards. The traveler was isolated from the passing scene, viewing it through plate glass, and could easily fall into a reverie, feeling that the train was stationary while the landscape rushed by. Moreover, as Stilgoe notes, the people in the landscape glimpsed from the train "struck passengers not as individuals but as a type."[11] From a passenger's point of view their aesthetic function was to animate the scene and provide human scale. Thus railway travel inculcated a taste for the picturesque view, while eliminating a landscape's details, distinct sounds, smells, and tastes, and preventing any direct contact between the traveler and local inhabitants.

Just as important, at stops along its route the railway delivered travelers to tourist sites *en masse*, creating new commercial opportunities. Niagara Falls provides perhaps the most striking example of how American tourism developed. John Sears and Elizabeth McKinsey have noted how, after the coming of the railway in the 1840s, Niagara was converted into a series of distinct view-points, each of which charged admission.[12] Some of these viewing positions were man-made walkways, towers, and boats, while others were natural rock formations or islands in the river. Their names and locations need not detain us, but it is worth noting that tourists were prepared to pay a good deal to see these vistas. One nineteenth-century tourist complained that it cost him eight dollars in fees to see the Falls during two days, which is the equivalent of more than one hundred dollars today.[13] The Niagara tourist experience consisted of paying admission to see a series of carefully constructed views. The structure of this encounter is that which Dean MacCannell described in *The Tourist*: the visitor is drawn to and recognizes a site by its iconographic representations and descriptions.[14] Off-site markers, such as the railway calendars and brochures about Niagara Falls, stimulate in the tourist a desire to see, recognize and visually appropriate a site. Yet paradoxically, as James knew, the off-site markers also annihilate surprise and wonder, almost in direct proportion to how magnificent a scene is supposed to be. One is not likely to experience intense disappointment in recognizing, based on photographs and descriptions, a regional building style, a plant, or an animal, such as a buffalo. But when an object is touted as one of the wonders of the earth, as is the case with Niagara Falls, extensive acquaintance with representations of the object can permit expectations to outrun reality, to such an extent that the object disappoints, and does not "live up to" what one had imagined.

This "egotistical sublime" was already a problem for tourists visiting Niagara before the American Civil War. The cataracts were extensively illustrated in books, universally praised, and visible far from the site in the form of giant panoramas and even a working scale model displayed in New York City.[15] Hawthorne and Margaret Fuller were both well aware of the dangers of great expectations, and both prescribed extensive acquaintance with the Falls, acquired over a period of days, as the only remedy, so that preconceptions could be overcome.[16] They understood that all five senses were necessary to apprehend Niagara, long before post-structuralism declared the inadequacy of language to describe such an object. Thus the characteristic goal of a sophisticated nineteenth-century tourist was not merely to recognize a site and check it off the list, but rather to live in close proximity to it, throwing aside temporary aids such as guide books and photographs. To serve this end many hotels were constructed near Niagara Falls, so that people literally could spend days or even weeks looking at it. In a characteristically romantic gesture, such tourists wanted to achieve a purity of vision unobstructed by prior expectations. For James, that kind of vision could be located historically, as pre-photographic, but through much of the nineteenth century many still believed in their ability to recover an innocence of vision, to see the familiar anew for themselves, an assertion that lies at the heart of the works of Ralph Waldo Emerson, Walt Whitman, and Henry David Thoreau. Such an ideology, combined with America's extensive rail network, created the conditions for a tourist industry. At Niagara it supplied a system of views along trails, bridges, towers, and other lookouts, designed to give each visitor the opportunity to reconstruct the object through many different encounters, until it finally achieved a three-dimensionality and complexity far greater than words and images could convey. While no doubt this was a commodified object being sold piecemeal to the public, that should by no means suggest a simplification of the object into a two-dimensional representation of itself. Rather, the whole effort of this kind of tourism was to exceed what could be represented or what had been pre-visualized, to penetrate to the "essence" of the object, moving beyond words toward sublimity. Just as the capitalist investment in stocks was expected to yield a surplus, the tourist's investment in a site was expected to yield more than its diagramatic representation. The ninetenth-century tourist's sublime was the psychological counterpart to a capitalist economy, in which an investment of time and effort was expected to pay a dividend.

Niagara was hardly an isolated example. Well before the Civil War Americans had invented a domestic grand tour, with the Falls as one of the major stops. It included the valley of the Hudson River, the Erie Canal, Niagara, the Great Lakes, Chicago, the Mississippi and Ohio Rivers, and a host of side trips to such places as Mamouth Cave and Virginia's Natural Bridge. In 1851 came word of a stupendous new natural wonder, Yosemite Valley, which was soon set aside as a park.[17] The process that had begun in the East would be transferred to the West.

II

The Grand Canyon was not often visited before the last decades of the nineteenth century, and did not become a national monument until 1908. As a tourist site it was constructed very much in accord with the Niagara model, in which a railway line advertised extensively and brought tourists to one location, where hotels and restaurants saw to bodily needs and a series of lookouts and trails were laid out to facilitate contemplation of a natural wonder. Before considering this site in detail, however, one must place the Canyon within the larger context of western travel.

The landscapes of the American West were ideally suited to the long perspective and large-scale vistas made available by railway cars. In the arid plateaus of the Rocky Mountains there was often little to see in the foreground, and few obstructing trees and bushes to prevent the traveler from seeing the mountains, buttes, and rugged country which were often in the distance.[18] The first transcontinental line, the Union Pacific, hired A.J. Russell to produce landscape images of scenes along their tracks, and during the year after the line opened many of these images, including that depicting the driving of the golden spike, became familiar to a wide public through popular magazines and lectures.[19] By the 1880s the American West began to rival the European grand tour as a tourist attraction to the wealthy elite. Books such as H. Hussey Vivian's *Notes of a Tour in America* described the western railroad and praised the unbroken views of scenery possible from the observation car located at the rear of the train.[20] A round trip from New York to San Francisco took six days each way and cost $300 (more for a sleeping berth), making the time and expense of such a journey somewhat less than a European sojourn, but it was still beyond the economic reach of most.[21]

Precisely during these decades, before most Americans could journey through the West, the first national parks were created, based largely upon the representations of these sites in paintings, photographs, and detailed descriptions submitted to Congress. The arrival of the railroad in the vicinity of a natural wonder simultaneously increased its value both as real estate and as a tourist site. The establishment of parks was explicitly intended to protect them from unscrupulous development, so that generations of tourists could appreciate them. The Grand Canyon was far from any centers of population and quite difficult of access throughout the nineteenth century, and therefore at first little pressure developed for its preservation. Relatively few travelers had the time or inclination to make a 112-mile round trip over a dirt road from the nearest rail-head at Williams. Once the Acheson, Topeka and Santa Fe built a spur to within a stone's throw of the South Rim and established a hotel there, however, tourists began to pour into the site, creating immediate demands for its protection.[22] It became a national park in 1919.

Today Grand Canyon still has the basic structure of a railroad tourist site. Virtually all of its hotels, campsites, souvenir shops and services are located within one mile of the railway depot. The most famous and often reproduced views of the Canyon are from a series of lookouts in this area, and here too begin the most popular trails into the Canyon from the South Rim. This layout implicitly assumes a relatively small public arriving by train, with leisure to hike several miles both east and west along the rim and also down into the Canyon. However, the Grand Canyon became a national icon in the era of automobiles, and as these proliferated the Park Service extended roads along the South Rim, paved them, and built parking lots near the major outlooks. This expansion had the effect of making it desirable to come by car, since few railroad passengers wanted to walk the long hot roads. Indeed, for many years railway service to the site was discontinued, though it recently has resumed. In 1996 the majority of the 5 million tourists drove along the southern rim on a two-lane road which itself offers few good views of the Canyon. There are turn-outs and parking areas at the major outlooks. The 30-mile length of this highway, the high temperatures of the summer months when most tourists visit, and the park service's principled decision not to erect more hotels and other conveniences outside the old core area, forces most tourists to rely extensively on their cars. The Grand Canyon is experienced as a series of views seen during brief stops along a congested highway. Because there is hotel and camping space for fewer than 4,000 people, most visitors drive in and out of the park each day during their stay. In the summer months 15,000 visitors a day is not unusual. One hundred years after the grand tour discovered the site, the walking paths along the rim still stretch for only a few miles, and they are often overrun. Likewise, the Kaibab and Bright Angel Trails are still the only easily accessible arteries for those wishing to go down into the Canyon. Both trails are congested during the summer.

Why do 5 million people have to traverse a single road to inadequate hotel space in a small area, when the Grand Canyon is more than 400 kilometers long? Because tourist demand has encountered an even stronger public demand that this site not be commercialized. As the result of an ecological concern for wilderness preservation, facilities inside the park have not expanded, although government projections are that by the year 2000 there will be 8 million people coming each year.[23] Because of the heavy traffic and the lack of adequate parking or accommodations, many visitors spend more time driving than looking. At Grand Canyon mass tourism based on the automobile has been imposed on a site designed for the grand tour. The automotive tourist expects to find services and conveniences spread out over a distended area. Instead, Grand Canyon Village concentrates all traffic at one site. Where the railway passenger did not decide where the train would stop, tourists in automobiles stop where it pleases them. Rather than watch a panorama roll by the train window, they elect to spend time looking at some views and then struggle through traffic to another lookout.

Yet the increased visitation is only one part of the change in perceptions. In the railway era the Grand Canyon, like Niagara Falls, invited reflection on human insignificance, and its system of views was designed to create an overwhelming impression of natural force. Today, much of the public sees the Grand Canyon through a cultural lens shaped by advanced technology. While ecologically-minded administrators want to designate up to 90 percent of the Canyon as wilderness, many visitors see the site as either man-made or requiring human improvements. Their characteristic questions, recorded by park staff, assume that human beings either dug out the Canyon or that they ought to improve it, so that it might be viewed more quickly and easily. There are repeated queries for directions to the (non-existent) road, elevator, funicular, train, bus, or trolley down to the bottom. Some even assume that the Canyon is man-made, produced by the Indians or a New Deal program. They ask: "What tools did they use?" Others want the Canyon to be lighted up at night, so that it can be seen round the clock.[24] I also overheard one young man declare it would be a great place for a rock concert. These remarks reveal a widespread belief that the Canyon, like any other tourist facility, ought to be improved. Visitors enjoy it in its undeveloped form, but they can see how much better it would be with more hotels, light shows, elevators, roads, tramways, and perhaps river boats and luxury hotels at the bottom. Where natural objects such as Niagara Falls and the Grand Canyon once were regarded as proof of the infinite power of the Creator, today the assumption of human omnipotence has become so common that it seems the natural world should always be made over for convenience.

The railway tourists to Grand Canyon at the turn of the century apparently seldom had such thoughts. Captain John Hance, one of the early promoters of the site, asked travelers to record their reactions after visiting the area, and in 1899 he published a book of *Personal Impressions of the Grand Canyon of the Colorado River*. Throughout this volume the word "sublime" is used continually, and in many cases the travelers make a point to say that elsewhere they have seen nothing remotely comparable. A visitor from New York typically wrote that, "After having visited all the noted places of both Europe and America, I have seen nothing to compare with the sublimity of the Grand Canyon." W. F. Cody (Buffalo Bill) visited with a party of thirteen, all of whom solemnly undersigned the statement that the Canyon was "too sublime for expression, too wonderful to behold without awe, and beyond all power of mortal description."[25] Most of those who wrote in Hance's book had not only looked over the rim of the Canyon, but also had been conducted by him down the Bright Angel Trail to the bottom of the inner canyon, spent the night there, and made the twelve-hour trek back up again. Their overnight accommodation was Phantom Ranch, which still remains the only hotel facility available at the bottom. This overnight, once the common and expected highlight of a visit to the Grand Canyon, has become the privilege of 100 people a day, less than one percent of the tourists. The trip to the bottom and back, whether on a mule or on foot, requires two full days. A recent traffic study found that the average time a tourist remains inside the park has diminished, from several days a century ago to less than twelve hours in 1990.[26] The construction and administration of the site makes long stays and detailed appreciation of the Grand Canyon nearly impossible for most visitors. Crowded along the South Rim, their experience can only be tantalizing and incomplete; they know that without a reservation at Phantom Ranch they can at best descend part way into the Canyon before returning to their car, to join the crawling procession leaving the park at dusk.

Given the crowds and the inconvenience of visiting Grand Canyon, many technological systems have been developed to deliver the "experience" of the park more quickly and easily. The most expensive of these is a flight over the park in a rented plane, with regular service from Phoenix, Las Vegas, or just outside the park. Papillon Grand Canyon Helicopters claims that each year it carries 100,000 customers. The de lux tour also includes a flight over Hoover Dam and an overnight stay at the bottom of the Canyon in the Havasupai Indian Village. Arizona Air flies in from Scotsdale, and includes in its package tour not only the flight over the Canyon but its representation in the popular IMAX theater, just outside the south entrance to the park near the airport. It boasts quadraphonic sound and a screen the size of a two-story house. The 34-minute film is screened once every hour. Its representations of flight over the Canyon and of the river raft trip are advertized as being more dizzying and terrifying than the real thing. Virtually all those who see the film also visit the park, but given the short time they have available, the theater makes it possible for them to avoid the hot climb into the Canyon and the considerable expense of a flight, while expanding their knowledge of the Canyon's appearance from different vantage points.

Significantly, IMAX theaters also have been erected at Niagara Falls and Yosemite. While such facilities could be located anywhere, their presence and commercial success near the entrances to these sites suggests that they fill a tourist's demand for more intensity and "authenticity." After seeing the actual park, screenings are presented not as substitutes for direct experience of the site, but as technological compressions of the many possible views into a tightly edited form. The slow acquisition of knowledge that once required several days is now packed into a half hour. The off-site marker has been transformed into an extension of the thing-in-itself. If the visual corollary of railway travel was landscape photography, such carefully framed views are no longer satisfying in the age of the airplane and automobile. The tourist now wants to be moved into the

landscape. The railway promoted static contemplation and the slow accumulation of impressions as one calmly passed a landscape. But the modern
tourist wants to penetrate into the site, by car, raft, hand-glider, helicopter,
or plane, or if that is impossible, through a film of one or more of these
experiences. The modern tourist seeks to enact his or her mastery over the
site, to make it an extension of human vision, and even goes so far as to
imagine that it is man-made. Where the nineteenth-century visitor sought
contemplative ease in order to allow Niagara and the Grand Canyon to
have an uplifting and transformative effect, the automotive tourist presses
for faster and more dynamic access. Instead of sublimity, what is wanted
are jump-cuts and a collage of novel sensations. The 1990s view of the
natural object expresses not a capitalist-investor's vision, but rather
consumerism. The contemporary tourist, viewing the landscape, thinks in
terms of speed and immediacy: the strongest possible experience in a
minimum of time. The grand tour proferred a leisurely banquet of the
senses; the post-modern tourist demands fast-food and a shot of adrenalin.

III

This shift in tourist sensibilities began with the rise of amusement parks.
These were appendages to the streetcar lines, which significantly were a
transitional form of transportation between the railway and the automobile.[27] By 1910 these trolley parks were situated on the outskirts of every
major American city, providing packaged forms of entertainment that
included the simulation of natural wonders and disasters. Patrons could
see re-enactments of the Chicago Fire, the San Francisco Earthquake, and
the Johnstown Flood, or they could journey to a grotto beneath Niagara
Falls. The front of the building that housed this exhibit at Coney Island
was "a great golden frame, showing the Falls as seen from the cliffs of the
Victoria Park Hotel on the Canadian side." The outside of the building
thus monumentalized landscape photography and painting. It employed
a well-established representation of a static view to lure customers to
experience a three-dimensional, dynamic display. Customers entered an
elevator which gave "the sensation of descending a great depth" as the air
grew "moist and chilly, and on stepping out at the bottom the visitors"
found "themselves in a cavern hewed out apparently from the solid rock.
Following the guide through tortuous slippery passages," they reached
"the back of the Falls at the base of a seething whirlpool, amid the dripping waters and deafening thunders of the giant cataract."[28] Such exhibits
prefigure the IMAX theaters of today. Both are presented as simulations of

Figure 2: Electrical engineer Charles Steinmetz and other cyclists apparently at Niagara Falls (trick photograph, c. 1900). *(Personal Collection, Author)*

natural wonders. Both cater to a mass audience. Both are technological creations. Both package experience and sell it as a commodity. Both employ a narrator (or guide) to interpret the meaning of their displays. Both claim for simulations the power to substitute for and amplify the meaning of the real. What changed between 1910 and 1990 was the power of technologies to deliver more powerful simulations, even as the number of tourists increased dramatically with each decade. Thus in Las Vegas a tourist can take a ride on a raft down an artificial Colorado River, in a $75-million amusement park at the Circus Circus casino. The ride is called "Grand Slam Canyon."

The construction of the tourist gaze is not simply the product of a chain of signification stretching from a site to an endless series of markers, inflected by advertisers.[29] Rather, that gaze is embedded in technological structures. It is shaped by the characteristic modes of transportation, the layout of facilities at the site, and the dominant forms of representation in each age, most obviously photography and film. A particular gaze emerged from the dominant tourist ideology of each time: conspicuous leisure in the age of capital accumulation and the grand tour; mass consumption in the age of capital disaccumulation and the short holiday. The mid-nineteenth-century tourist entered a setting designed to dramatize nature in terms of the sublime. In contrast, the late twentieth-century tourist seeks not transcendence but powerful sensations, not stasis but action, not the contemplation of natural forces but the expression of human power. Even the ecological view of a natural wonder is built upon the realization that human beings can destroy nature, and the Grand Canyon increasingly is apprehended as a fragile physical and biological system threatened by pollution, acid rain, hydroelectric dams, and the presence of the tourists themselves. From the moment that visitors began to frequent Niagara Falls, which is to say from the inception of American tourism, changing technologies have revised the narratives which shape the tourist's gaze. Niagara Falls and the Grand Canyon still represent America to today's tourist, but these icons are part of much different systems of meaning than they were a century ago.

I

For over a century Americans have wrestled with an intriguing but fundamentally misguided idea, that the conquest of the western United States was the central experience that defined the nation. This "frontier thesis" was first articulated as an essay in 1894 by a young historian, Frederick Jackson Turner, and battalions of later historians have never entirely dislodged his theory, which remains popular with the public.[1] Curiously, the frontier Turner described contained little technology, as he undervalued new tools and machines. For him, the West was a moving line of settlement with a low density of population, less than two per square mile. Such a definition assumes that the West was settled piecemeal by individuals, moving out on their own to confront wild nature. As Richard White points out, "Turner's farmers were peaceful; they overcame a wilderness; Indians figured only peripherally in this story."[2] For Turner, the West was essentially vacant, an abstract space awaiting transformation into the American landscape. He declared, "The existence of an area of free land, its continuous recession, and the advance of American settlement westward, explain American development."[3] As many commentators have pointed out, the land was not free, but had been inhabited for millennia. And only the Anglo-American settlers expanded from east to west. The Spanish came from the south, the French from the north, and Russians, Chinese and Japanese from the west. These and many other deficiencies have been amply documented.[4]

Turner's lack of interest in technology has been less often noted, however. It is suggested by the single word "continuous," as though the settlement of the West had been a single uninterrupted movement by individuals. It took almost 200 years for American colonists to settle the area east of the Appalachians, but little more than two generations to move from Pittsburgh to San Francisco. The leap across the continent was possible because Anglo-Americans developed new and powerful technologies. In the imagination the West often seems pre-technological, as if it alone among the regions remained an elemental battleground between man and nature. The "new" western history often takes Turner to task for being insensitive to Native Americans or for concentrating too much on male experience, but it less common to find critiques which suggest that he ignored the most powerful force in the settlement of the West, namely the technologies that white Americans had at their command. Where the seventeenth-century settlers of the Atlantic seaboard relied on oxen to drag their produce to market, the vast region west of the 100th meridian was crossed by railroads that brought miners canned food, steel tools, dynamite, and processing machinery. To farmers they brought barbed wire, mechanical reapers, easy-to-assemble windmills, glass for windows, and an array of household utensils. Most important of all, railroads provided an outlet for the products of the West, carrying ore, timber, cattle, and farm produce to market. Settlement was not evenly spread along a "continuous moving line." Rather, settlement focused on the railroad, the irrigation ditch, and, after the 1880s, the electric utility line. Electricity and the railroad are more useful to the understanding the region's history than the Turner thesis.

One basic problem is to decide the West's boundaries. Almost any region of the present United States, such as western Massachusetts or Kentucky, was once someone's Wild West. Any definition must acknowledge climatic conditions, particularly the cycles of rainfall, and the arbitrary political boundaries with Canada and Mexico, which slice across mountains and rivers. Here, the West will be defined as the area from the high plains of the Dakotas, Nebraska and west Texas to the Pacific. The time period examined will be from 1880 to 1940, the years when electrification spread over the region. The transcontinental railroad had linked California with the rest of the country in 1869, and by 1880 Denver and San Francisco had already mushroomed into what Gunther Barth rightly calls "instant cities."[5] By then the military conquest of the area was largely over, and the spread of American civilization at the expense of Native-American and Mexican cultures was well underway, as the interior of the West was being settled from both the east and the Pacific coast.

Settlement of the mountainous west and the Pacific coast did not simply repeat the pattern of earlier migrations, in which agriculture played a leading role. Americans were drawn out to the great plains by cheap land, much of it distributed under legislation such as the Homestead Act. But after 1880 the availability of electricity transformed patterns of migration and settlement in the West, which would not have developed in the same way without it. In terms of electrification, there were four distinct wests, which became important in a chronological sequence unlike that of other regions. First came the mining west, itself a technological phenomenon dependent on railroads and other forms of advanced technology. This stimulated the urban west, particularly Denver and San Francisco. Third (not at the start as Turner would have it) came the agricultural west, itself split by ranching and farming interests, and, fourth, the regions that remained became significant precisely because they were undeveloped, or wilderness. Electrification meant something different for each one of these four, and in combination they defined a region quite unlike the South, East, or Middle West. Indeed, the development of the electrical grid proved to be inseparable from many of the central conflicts of western history: federal versus local interests, industrialization versus preservation, agriculture versus ranching, conflicts over water rights, tribal rights versus plans for development, and private versus public ownership of utilities. Electrification was hardly a "natural" or an "inevitable" process, but a social construction whose form was determined by these contradictions.

II

Mining was central to the region's early development, as discoveries of gold, silver, lead, and copper drew many of the first settlers and investors to Colorado, California, South Dakota, Montana, and Nevada. Mines created many of the early urban centers: Virginia City, Nevada; Butte, Montana; Deadwood, South Dakota; Leadville, Colorado. They stimulated the regional economies and made many railroad lines economically feasible. They attracted outside investment. Without the income they created and the settlers they attracted, development would have been slow. The Comstock Lode in Virginia City alone produced $300 million in a twenty-year period, creating the boom town that Mark Twain described in *Roughing It*. The first large cities, Denver and San Francisco, drew much of their wealth from proximity to gold and silver mines. Both cities provided machinery, goods, and services to the mining communities, such as the expensive foods and clothing that miners preferred after they had made a strike.[6] Indeed, mining towns were often thrown up in a season, using imported materials. Usually located in regions without farms, they could only specialize in extracting ore because virtually everything necessary to daily life came from elsewhere. Denver and San Francisco served as the transhipment and banking centers for many of these communities.

While the importance of the railroad is evident, electricity was used to improve the productivity of the mines. Its introduction thus stimulated the economy of the entire region. Until the last decades of the nineteenth century inadequate power and light were obstacles to rapid exploitation of claims. The advantages of good lighting in mines were so obvious that electricity was employed immediately. In the spring of 1879, only one year after the Brush Company began to manufacture arc lights in Cleveland and more than half a year before Edison's first public demonstration of the incandescent lamp, an arc-light system already had been installed in a Yuba County gold mine. The *Nevada City Transcript* reported: "The first electric light ever introduced in a mining claim was placed on the Deer Creek Claim of the Excelsior Water Company. ... A 12,000 candle-power Brush machine was put in operation and three lights of 3,000 candle-power were placed in prominent positions on the claim. ... Although the night was very dark, the lights shed a brilliant light around and enabled the miners to work as readily as during the day."[7] As the paper noted, night operation of the mine permitted it nearly to double the daily output. *The Mining and Scientific Press* picked up the story and many others soon began to install their own generating plants.

While mines for precious metals quickly exploited electric lighting, all mines needed industrial power to haul ore to the surface, to run lifts, and to operate crushing plants. Producing base metals and coal in any quantity absolutely required a constant energy source. Traditionally, power had been provided by water mills or steam engines. But in the arid West water was seldom available close to the mine. Mules or men also could haul ore to the surface, but both were slow and expensive. Steam engines were better, but they required a constant supply of clean water and large amounts of wood or coal. Furthermore, steam engines were not easily portable, and could not be quickly moved when a mine opened new shafts or otherwise changed its power requirements. The advantage of electric power, of course, was that it could be generated from a distant rushing stream or from a coal-fired steam generator located at a railhead, and then transmitted to where it was required. The company no longer needed to haul a steam engine, water, and fuel to the mine-head, and electric motors could be placed at any location in the shafts and tunnels. As early as 1892 one of Edison's lieutenants, John Kreusi, made field reports which showed that an Edison Electric Mine Locomotive could haul 21 loaded cars along a 4000-foot tunnel at the Loyal Hanna Mine. The engine was operated by a former mule driver who "never saw a piece of electrical apparatus until a month ago" yet learned to operate it with only a few days of instruction.[8] Such electric locomotives proved to be cheaper than mules in hauling coal to the surface, cutting costs in half. Other new electrical technologies also began to appear, including electric drills and electric coal cutters, though the latter had not been perfected and temporarily had to be withdrawn from the market. General Electric sold generating plants and equipment to mines throughout the West, speeding the pace of regional production and development. For example, it equipped both the mills of the Homestake Mining Company in South Dakota and the first completely electrified open-pit coal mine, in Colstrip, Montana in 1925.[9]

The use of advanced electrical technology transformed the miner's work. As mules and men were replaced by small electric locomotives, drills, and cutters, unskilled jobs disappeared and new demands on the remaining workers changed the nature of the work. The rapid elimination of manual labor gave greater control of the work process to management, and was one of the causes of union militancy in the West, where the Wobblies were particularly strong, and where a series of violent strikes occurred between 1890 and 1920.

III

One way for an upstart mining town to show its advanced state of civilization and its equality with the East was to install electric lamps on the main street. Electricity could serve symbolic purposes virtually anywhere. In 1879 the Palace Hotel in San Francisco installed arc lights in its public rooms. Long before most homes had electricity, expensive brothels displayed their opulence with cut crystal chandeliers using electric bulbs. Nor was the symbolic use of lighting a late development. At one of the first demonstrations of electric lighting, in 1871, Father Joseph Neri, a scientist on the faculty of St. Ignatius College, San Francisco, placed an electric arc light in a college window to honor the silver jubilee of Pope Pius the IX.[10] While Father Neri made a number of other public demonstrations, including a searchlight that was reportedly visible for more than one hundred miles, his work did not lead to the formation of a local electrical manufacturing industry. Instead, generating equipment and lighting apparatus were imported from the East and from Europe. In 1878 one of the

San Francisco *Chronicle's* editors brought back two arc lights from the Paris Exposition and had them permanently installed outside the newspaper's new building.[11] Such a display was then a rarity and caused a sensation. A year later in 1879 San Francisco had one of the first electrical utilities in the world, The California Electric Light Company.

As this example suggests, electricity came to the West more quickly than to most other sections of the nation despite the almost complete lack of an industrial base. The vast electrical grid which covers the West today spread from a few isolated electrical plants set up in the mines and main cities. The *Chronicle* used steam from its heating system to drive a generator in 1878; four years later Edison's incandescent lighting system came to San Francisco, in the same year that Edison first operated a utility in New York City. One George S. Ladd purchased and installed an isolated plant to illuminate sixty lamps in his business. In the middle of the decade Denver adopted the then popular idea of erecting powerful arc lights on two hundred-foot towers to light up the city.[12] By the 1890s, however, the Edison system, modified to work with alternating current, was rapidly becoming the norm throughout the West.

As system design became standardized, however, the more difficult question of social control became central. Should utilities be public or private? In Tacoma and Seattle the answer was "public." Seattle held an election on the issue in 1902 "to decide if the city power plant should be built" to compete with the monopoly of Seattle Electric, and voted in favor seven to one.[13] Competition provided lower electric rates than the average, and Tacoma and Seattle rapidly achieved some of the highest levels of home electrical consumption in the United States. By contrast, in California private enterprise triumphed. New York financiers arranged the formation of the Bay Counties Power Company, and underwrote the construction of a 140-mile transmission line, then the world's longest, from the mountains to Oakland. In 1905 another group of New York investors merged several smaller companies into Pacific Gas and Electric (PG & E). As Thomas P. Hughes shows in *Networks of Power*, by 1914 PG & E controlled an integrated regional network, covering 30 counties, or most of central California, and supplied 1.3 million people, then nearly half the state's population. In subsequent years the company absorbed its remaining competition, and vigorously opposed a state Water and Power Act that would have required the state government to distribute electricity generated from public dams.[14] In Los Angeles, private power also triumphed, but the capital came from local sources. The railway czar, Harry E. Huntington, owned most of the stock in Pacific Light & Power, which by the early 1920s had merged with its two largest competitors. The service such large companies offered was comprehensive, but their rates at times were higher than in municipal systems. Perhaps for this reason, a 1925 survey conducted by the National Electric Light Association showed that California distinctly lagged behind other regions in the extent of the use of electricity in the home.[15]

These were the two extremes in utility development, power as a public service, and power as a profitable commodity. The two systems did not peacefully coexist. The generally lower rates of municipal systems were a constant embarrassment to private companies, who attacked public power as a dangerous form of socialism. At the center of this conflict one often found the question of dam construction and competition for the right to buy inexpensive power generated at federal dams. As President Theodore Roosevelt put it, "corporations are acting with foresight, singleness of purpose and vigor to control the water power of the country."[16] Progressives attempted both to restrict private acquisition of irrigable land and to propose legislation that would institute a "multiple-purpose policy" that combined flood control, forestry, irrigation, river navigation, and electrical generation in one comprehensive policy. But the Water Power Act of 1920 was only a compromise, which strengthened government supervision of utilities and abolished long-term private leases on dam sites. It still left hydroelectric development primarily in private hands.

During the 1920s the progressive minority in Washington DC led by Frank Norris, Robert LaFollette, Burton K. Wheeler, Thomas J. Walsh, and Clarence Dill protested the concentration of the electric power industry in holding companies, and after the Teapot Dome Scandal they sponsored two investigations, the first on whether or not a power trust existed, the second on the extensive public relations activities of the utility industry. Together their work prepared the way for New Deal utility legislation and final adoption of the "multiple purpose" policy.[17] At the end of World War II, however, private utilities still battled public power companies for the right to purchase electricity from federal dams. At the same time they opposed a plan to turn the Columbia River Valley into a Tennessee Valley Authority for the Pacific Northwest.[18] Similarly, in California, "Pacific Gas and Electric Company managed to block every congressional appropriation for government-owned transmission lines. When the generating plant at Shasta Dam began to produce power in 1944, Secretary of the Interior Ickes had no choice but to negotiate a five year agreement for the sale of all of this power to PG and E."[19]

IV

The value of private utility monopolies was particularly high in the West, because its industry did not develop in the same way as on the east coast, but relied heavily on electric heat and power. Dependable electric motors were not available until 1885. Before then a few western manufacturers had invested in plants located along swift moving streams or adopted steam power. The best early motors were made by Frank Sprague and marketed by the Edison companies. To demonstrate the industrial uses of electrical motors, which in 1889 supplied less than 1 percent of American factory horse-power, the Sprague Electric Company installed a hydroelectric system on the Feather River in California.[20] This was one of the first industrial developments based on electric power anywhere in the nation. The generating plant supplied power to fourteen different manufacturers nearby, including a sawmill and an ore-crushing plant. Sprague's motors also drove electric drills, blowers, and hoists. More importantly, Sprague demonstrated that electricity could compete with water-powered mills, which usually lost between 25 and 40 percent of their power due to friction in turning the shafts and belts. Tests showed that at the Feather River installation the electrical system was at least as efficient, which meant that a manufacturer operating on a water wheel would need no additional flow of water to convert the same facility to electricity.[21]

Sprague's system had one serious disadvantage, however. It worked on direct current, which could not be transmitted over distances of more than a few miles. Alternating current, which was just coming into commercial use in 1889, could eliminate this difficulty, so that factories no longer would be compelled to locate near an electricity-generating station. Instead, they could be located more conveniently in town or along a rail line. A dramatic example of how the success of alternating current affected factory location came in 1893, at Folsom, California. There, Horatio Livermore had planned an industrial city based on the model of Lowell, the Massachusetts mill-town which had successfully exploited water power since the 1830s, and which had long been famous as a center of New England's industrial development.[22] Livermore completed a dam on the American River in 1893, but abandoned plans to build manufacturing plants of the sort then still commonly found in the East, driven by water turbines. Instead, he decided to generate electricity and transmit it twenty miles to Sacramento, using the new system of alternating current that recently had been successfully demonstrated in Germany. Using General Electric equipment, in 1895 the Folsom plant transmitted 3,000 kilowatts to Sacramento, to light its streets, run the streetcar line, and power local industries.

The American River Project was just the beginning. While a few small power stations could be run on coal then imported from Australia, or oil, which was soon discovered in Los Angeles, the most promising energy source in the West was water power. The success of alternating current made it clear that the rivers of the high mountain country could supply inexpensive energy to cities, creating the basis for western industrialization. Even as the Folsom plant was being completed, Westinghouse alternating-current equipment was being installed in another Californian hydroelectric power plant. The new technology radically changed the requirements for factory location. The West would have few of the small mill-towns that dotted the southern uplands and eastern fall line, where earlier factories had to be built in order to use water power.[23] Virtually from the start, industries in the West could be located anywhere electrical lines could reach, and they usually were concentrated in cities. Thus electrification early contributed to the striking concentration of the West's sparse population in cities, helping to make it the most heavily urbanized region in the United States today, with 84 percent of its people in cities.[24]

The availability of electric power at the inception of industrialization on the west coast stimulated adoption of the latest electrical technologies. For example, in 1900 General Electric installed electric motors in the California Cotton Mills in Oakland, putting it in the vanguard of American mills. In Sacramento, the Southern Pacific was one of the first railways to install an electrical shop to service its engines. By 1902 the western market looked so lucrative to Westinghouse that it had located five of its twenty branch offices in San Francisco, Los Angeles, Seattle, Salt Lake City, and Denver. The continued construction of large hydroelectric dams throughout the West (discussed below) attracted corporations that required large amounts of energy, such as smelters, aluminum plants, and chemical works.

Before American cities could expand in response to industrial growth, they needed improved mass transportation. The walking city could extend only a mile or two from its center before becoming unwieldy. The horsecar extended the radius of the circle, making a city of twelve square miles feasible, but the cable car and the trolley made possible residential areas covering eighty square miles, nearly seven times as large.[25] In 1873 San Francisco was the first city to establish an extensive cable-car system, retained to this day. Because of its steep hills, such a system is preferable, but everywhere else trolleys conquered the West. Denver briefly installed an early streetcar system designed by a Professor Short from its own university, then adopted cable cars in the 1880s, and finally shifted back to the electric streetcar again, once it had been perfected by Sprague in 1888.[26] However, it was Los Angeles which adopted the trolley most wholeheartedly. It had 145 miles of track already in 1890, suggesting that its present dependence on the motor car was hardly inevitable. The Los Angeles system was as extensive as that in St. Louis or Pittsburgh, then larger cities, whose lines carried four or five times as many passengers annually.[27] The Los Angeles trolley system was designed and controlled by real estate speculators to make otherwise remote suburban lots attractive. Again, Henry Huntington was a central figure, as his Pacific Electric Railway Company connected 42 communities in a 35-mile circle around Los Angeles. Likewise, in Seattle virtually all of the suburban expansion between 1900 and 1920 was along its streetcar lines.[28]

In 1900 these streetcar and trolley networks were essential to the city's daily functioning. In hilly San Francisco the cable cars provided no less than an average of 340 rides per inhabitant per year, while Los Angeles and Denver hovered closer to the national average of 250. Even cities of less than 25,000 had good systems, including Great Falls, Montana (12 miles of track; 939,000 passengers), San Diego (16 miles; two million passengers), and Ogden, Utah, (11 miles; 860,000 passengers). In 1900, California alone had 35 electric streetcar systems carrying 182 million passengers, or 123 rides per state resident annually. Far fewer rode horses.[29]

Streetcars were not merely urban phenomena. Compared to steam railroads, they were less expensive to construct, able to climb steeper grades, capable of starting and stopping with less loss of time, and able to charge lower fares. The interurban trolleys served many western outlying regions. They carried children to school in parts of Utah, Idaho, and northern California; everywhere they made package deliveries to small towns and crossroads, and carried farm produce into the city. Some lines were used extensively for freight, and one in the Yakima Valley was constructed primarily to haul fruit. Indeed, the availability of the interurban often stimulated agriculture. One contemporary observed that "In Southern California, the trolley lines not only serve the farmers but conduct an information bureau, where they furnish information to prospective settlers in order to encourage them to take farms along the lines."[30] Thus the trolley and interurban together did not merely connect city and country, but stimulated new settlement patterns and agricultural development. Interurban lines often sold electricity to small towns and farming communities along their routes. In 1902 ten western interurban trolley lines sold street and private lighting, usually in remote areas which utilities could not profitably serve.[31]

Interurban systems also played an important role in stimulating local tourism, making the countryside accessible to ordinary people in the evening and on weekends. Los Angeles' residents could choose from three quite different interurban rides on the Pacific Electric Railway. They could take the Orange Empire Trolley Trip though the orchards and vineyards of the "world's greatest fruit domain." A more dramatic alternative was the climb up Mt. Lowe, "to near the bald top 6,100 feet in Cloudland." Or they could make a combined trolley and ship excursion to Santa Catalina Island, "enveloped in soft summer air and set in a turquoise sea. . ."[32] Once atop a mountain, on an island, or in a small town, the traveler often could stay the night, as trolleys opened up tourism to a new social class. Local residents near interurban lines saw the chance to rent out rooms in "tourist homes," and to open cafes, restaurants, and other small businesses. Thus, as streetcars linked rural people to the urban economy, they simultaneously made country life available to city people on weekends. Now anyone with a quarter and a free afternoon could make a long trip into the countryside.

Overall, there were fewer accidents with streetcars than with most other forms of transport. Despite fears of electrocution expressed by early travelers, very few died in this way. Most accidents occurred when pedestrians were hit by the cars. However, a small number of spectacular accidents undercut the streetcar's reputation for safety. One of the worst in the nation happened on July 4, 1900 at Tacoma Washington, where 36 people were killed and 60 injured when a crowded car jumped the track while crossing a bridge. The car plunged one hundred feet into a gulch.[33] Such serious accidents were quite rare, however, and overall the trolley was far safer than the automobile. Nevertheless, the trolley systems rapidly disappeared, in part due to competition from automobiles, in part because as instruments of land speculation they had fulfilled their purpose. The vastness of the West better suited the automobile for long-distance travel, but the trolley rather than the freeway might have served as the basis for western city planning. Only San Francisco retains a vestige of this era, because its cable cars, well suited to the steep inclines, have become icons of the city.

V

Seen from the white man's point of view, agriculture bloomed late in much of the West. First came the gold rushes and instant cities, and in the first generation neither the mining camps nor the early settlements could feed themselves from local foodstuffs. Instead, they imported provisions from the East. But if agriculture came late to much of the region, it arrived in an advanced form, incorporating irrigation and electrical technologies far more rapidly than the rest of the nation. For the United States as a whole, rural electrification came slowly. As late as 1935, only one farm in nine had electricity, compared to levels of close to 100 percent in Holland and 90 percent in France, Germany, and New Zealand. The West, one might assume, would be the least electrified region of all, because of the vast distances between power stations and farms. This was true in the sparsely-populated interior. In the western states of Texas, Oklahoma, the Dakotas, Wyoming and Montana less than 5 percent of the farmers had electricity. The few cattle ranchers or wheat farmers who did often generated it themselves using wind power or gasoline engines. Little changed until the federal government created the Rural Electrification Administration (REA) in 1936, which provided money and expertise to local cooperatives.[34] These states were vigorous participants in the REA. In contrast, California, Utah, and Washington had before 1935 achieved the highest levels of private rural electrification in the country. Fifty percent of their farms were electrified, five times the national average. Not incidentally, these were often irrigated lands that used electric pumps to distribute water. (While steam engines could also be used for pumping, they had the same disadvantages already noted with regard to mining.) In areas requiring irrigation—Utah, Nevada, Arizona, California—electricity did not merely improve farming, it was a precondition for it, except in a few easily irrigated places that were developed first.

The far West reversed the historical pattern that Turner posited in his famous essay, in which agriculture preceded urbanization and industry. The western plow followed eastern technology. As developed by Anglo-Americans, California, Nevada, Arizona, New Mexico, and Colorado were urban and mining centers that first imported food, and then engaged in self-conscious programs of rural development. As a character in Ednah Aiken's *The River* explained, "Irrigation is the creed of the West. Gold brought people to this country; water, scientifically applied, will keep them here."[35] The same hydroelectric dams that provided electrical power also held the necessary water in their reservoirs. The first combination irrigation–electrification project came already in the 1880s, when a self-educated engineer, George Chaffee, created the California communities of Etiwanda and Ontario. From the experience gained in such projects came more grandiose designs, including the taming of the Colorado River. In 1907 Laguna Dam was completed to water and electrify a stretch of

Figure 3: Hoover Dam under construction *(Smithsonian Institution)*

desert near the Mexican border that had been renamed "The Imperial Valley" of California.

Most western irrigation was accomplished with federal help, and today it waters 42,000,000 acres, representing 85 percent of all irrigated farms in the nation. Such projects made possible Arizona's cotton crop and California's enormous production of fruits and vegetables. Electricity is essential to make California the nation's most profitable agricultural state.[36] This was achieved without the REA. Of its 985 active cooperatives in 1979, only five were in California compared to 34 in South Dakota or 25 in Montana. Put another way, irrigation meant high energy consumption in concentrated areas, creating an attractive market for private utilities before 1930. In contrast, wheat farming and cattle ranching meant low demand for electricity spread over a dispersed area, creating a market so unattractive that only federal intervention after 1935 could bring power into these regions.

The most dramatic forms of federal intervention were the great dam building projects: Hoover Dam, Bonneville Dam, Grand Coulee Dam, and many others. Collectively these served many purposes, including flood control, irrigation, and recreation, but all were for electric power generation. The division of water and electricity from such projects became a matter of controversy, particularly between California and Arizona, which several times went to court and in 1934 called out the National Guard to halt federal dam construction.[37] Perhaps the most controversial of these projects was what came to be called Hoover Dam on the Colorado River, authorized by the Coolidge administration in 1928. Built to serve seven states with electricity, the dam permitted irrigation of one million acres in Arizona and California. Even after the dam was completed in 1936, however, Arizona opposed the division of water in court, because California was to receive almost twice as much. Electricity from the dam was also unequally divided. Arizona only in 1944 reluctantly agreed to the Colorado River Compact, which the other six states had signed in 1922.[38] If dam building in the parched Colorado region generated conflict over water rights, the Bonneville Dam on the Columbia River sparked a battle over electricity rights between public and private companies. These conflicts over water and electricity point to a still developing crisis in the West, as the water supply is as scarce today as at mid-century, while the population continues to grow.

Furthermore, as Donald Worster has argued in *Rivers of Empire*, the vast water projects have distorted the social and political structure of western states. In the nineteenth century the Mormons used church control of irrigation systems to impose social control. Economic rather than religious dominance was the goal elsewhere. As the scale of projects increased, however, local capital was unable to underwrite the costs, and western elites turned to Washington for assistance. Thus came the 1902 Newlands legislation which created the agency later called the Federal Reclamation Bureau. The idea was to finance irrigation through sale of land to homesteaders. Where Turner supposed that western lands were natural and free, irrigated lands turned out to be expensive cultural inventions. In practice most of the land was concentrated in the hands of a few owners, as for example near the Theodore Roosevelt Dam. When it became evident that such schemes would not be self-financing, more direct federal subsidies were sought, leading to the contruction of the Hoover Dam and a string of other dams up and down the Colorado River, and even a man-made river that had to be pumped over the mountains to California.[39] Control of water and electricity gave enormous power to political and economic elites. Just as trolley lines lined the pockets of real estate speculators, so too, as the film *Chinatown* suggested, control of irrigation water became a matter of political intrigue and fortune-building. When the Okies came to California from the Dust Bowl they found not the family farming that they knew, but a vast agribusiness which employed them as day laborers in large vineyards, orchards, and lettuce fields. As already had been the case with hydroelectric power for industry, with advanced electrical technologies for mines, and with the privately controlled street-railway interests, the irrigation projects which made the West the most rapidly developing area of the United States also concentrated wealth and power.

For most Americans, however, there was a huge gap between the pastoral West of the imagination and the actuality of capitalist irrigation farming that served a largely urban population. Many water projects were expected to bolster family homesteading in the desert, and migrants from the Dust Bowl often did not realize that most of the irrigible land was concentrated in large holdings. As Woody Guthrie put it in a famous song, hundreds of thousands of people in the Dust Bowl were "beatin' the hot and dusty way to the California line," under the illusion that California was a "garden of Eden," only to find that this was a paradise only for those with land and money.[40]

Many who went to California understood western development in the rhetoric of American exceptionalism, in which white men had "won" the West and transformed it into a paradise. Most Americans in the 1930s still shared a nineteenth-century vision of America, as a garden where culture and nature intermingled, in what Leo Marx has called a "middle land-scape" between raw wilderness and civilization.[41] Ever since Crèvecoeur's *Letters of an American Farmer* described such a pastoral landscape in the late eighteenth century, the region's imaginary location has shifted westward in retreat from industrialization. The gap constantly widened between the rhetorical evocation of the West as open land of equal opportunity and its actual hardships for small farmers, field hands, immigrants, and adventurers. Guthrie was trying to warn the migrants of the 1930s that Turner's homesteading West no longer existed. Indeed, it never had.

VI

If the electrification of the West had little to do with increased economic democracy, it had even less to do with preserving ecological balance. A region of unspoiled natural beauty a century ago has been transformed by strip mining, dams, urban sprawl, and irrigation. Whatever the many advantages of electricity, the increasing American demand for it has required far more than can come from hydroelectric dams. By the middle of the 1970s electrical production was a major source of pollution, as 45 percent of all electricity came from coal, 18 percent from natural gas, and 17 percent from oil, all raw materials found abundantly in the West.[42] Nor have all western sources been commercially developed, for large reserves of tar sands and shale oil deposits cannot be profitably exploited. Reserves of oil shale exceed 2 trillion barrels of oil, "about four times the entire crude oil reserves of the world" as estimated in 1980. The federal government owns three fourths of these reserves.[43] The oil and coal that have been extracted already have caused environmental damage, protests from American Indians who wish to preserve their homelands, and conflicts between those who want development and others who wish to preserve the wilderness.

Holding large portions of the West inviolate against encroaching civilization was an American dream well before Turner made his famous address. But the West's development bore little resemblance to the fable he gave his contemporaries, in which agriculture preceded industry, and culture followed the plow. One might say, rather, that in the West civilization followed the railroad, the mine, the high-tension line, the streetcar, and the irrigation ditch. All of these technologies required capital, and western economic growth was accelerated not so much by lone pioneers as by eastern investment, principally in railroads, dams, and electrical equipment. Without question a West lacking electrical power would have developed more slowly and differently. Electrification was essential to rapid exploitation of many mines, to pumping the water needed for irrigation, to early urban transportation, to suburban land development, to efficient utilization of the power of streams in the high mountains, and to the establishment of many high-energy industries.

But if Turner's famous thesis contradicted in many ways the actual development of the West, it served as the shorthand expression of a powerful ideology in which the United States was identified with Nature. This ideology can be traced to the philosophy of John Locke and to early religious and utopian conceptions of the New World, and it found full expression in the rhetoric of the Jacksonian era.[44] Yet even as this identification with nature became an unquestioned facet of national identity, industrialization was rapidly destroying the wilderness. Since there was little wild nature left in the South and East by 1860, citizens took a strong interest in preserving portions of the West as last vestiges of the empty space they understood America to have been. The process began with setting aside Yosemite and Yellowstone as parks, and developed into a national system of parks and monuments.[45]

As environmental and wilderness groups such as the Sierra Club grew more powerful, development of western irrigation and hydroelectric power became more diffcult. Environmentalists were appalled by much of what electrification meant, and tried to preserve a portion of the West untouched by high-tension lines, strip mining, oil wells, dams, or irrigation projects. The Sierra Club was founded as a reaction to the Hetch-Hetchy Dam project, which flooded a spectacular valley adjacent to Yosemite. Since the federal government today owns no less than 60 percent of the total area of the West, environmentalists have often operated through Washington.[46] The federal government became the mediator between preservationists and those pushing for development. The conflict began in the first decade of the twentieth century, as Theodore Roosevelt's administration enlarged the national forests. Such federal land remained open to exploitation, for grazing, logging, and mineral extraction. To prevent such abuses environmentalists sought to create wilderness areas, which unlike national parks or national forests could not be used for tourism, mining, or cattle grazing.

One fundamental premise of most environmentalists has been an opposition between nature and culture. A larger cultural result of electrification, however, has been the erosion of the distinction between these two terms. Electrification has fundamentally transformed the western landscape. Hydroelectric dams have made most of the West's large lakes; high-tension lines march across its mountains; water projects have made desert into farmland. All of these changes, occurring in a single lifetime, have become powerful metaphors for progress, making other extravagant projects and transformations seem realizable.

Nor are these massive geographical changes all. Electricity has always had symbolic uses. Even the first light put up by Father Neri was intended as a spectacle, and urban spectacles and amusements have relied heavily on electricity. As early as 1879 San Francisco celebrated the visit of Ullysses S. Grant by erecting a display of arc lights. Electric lighting provided a visual excitement to the city's landscape, transforming and editing its appearance. The memory and meaning of San Francisco is inseparable from lights along the bay and on the Golden Gate Bridge. What would Las Vegas or Disneyland be without lighting? Such sites originated in the amusement parks at the ends of streetcar lines, which created them to stimulate business during evenings and weekends. And because most early traction companies also produced their own electricity, they lavished lighting on their parks to make them more attractive.

These lessons were not lost on designers of world's fairs, who incorporated spectacular lighting into every major fair after the Columbian Exposition of 1893–1894. By the time that San Francisco held the Panama–Pacific Exposition of 1915, lighting had become a major component of fair design. In contrast to earlier fairs which had used many individual bulbs to outline the architectural features of major buildings, the Panama–Pacific Exposition took advantage of a new electrical technology —floodlighting—to create an entirely new visual experience.[47] The new lighting enabled organizers to erase the sharp distinction between day and night. As night fell, natural light was gradually supplemented with electric light, so that the towers of the fair were never in darkness. A building that was white by day, and tinged with rose at sunset gradually became a deeper red. Such treatment was particularly effective with the fair's central symbol, the "Tower of Jewels," a 435-foot structure encrusted with 100,000 hand-cut crystals, called Nova-gems, which refracted the powerful spotlights in shimmering rainbows of color. Appropriate lighting effects marked special occasions: all the buildings were colored green for St. Patrick's Day, and on the anniversary of the San Francisco Earthquake "the whole of the buildings were bathed in a fiery red, and, by cunning manipulation of the lights and the introduction of effects… a strikingly realistic illustration of the destruction of the Tower of Jewels by fire was presented."[48]

Other imitations of nature included artificial sunsets and a false aurora borealis, both staged using banks of searchlights on naval ships in San

Francisco Bay. These harbor facilities, combined with appropriate sound effects from a hissing and belching locomotive at the fair created extravagant fire-less fireworks, displays that "brought in more admissions at the San Francisco Exposition than any other attraction." One fountain showed the creation of the Earth out of the void. In another courtyard, "were two fountains symbolizing the rising and the setting sun respectively."[49]

In short, as early as 1915 there were few conceits that could not be realized with lighting effects. Lighting was used to make an illusory landscape, to erase the sense of sharp distinction between the natural and the man-made. Light thus created an impossible middle realm between nature and culture, harmonizing them in spectacular demonstrations that held spectators spellbound.[50] The western penchant for tall tales and extravagances found a new outlet, and was carried to extremes in amusement parks, in the electric signs along its urban highways, and in the apotheosis of "the strip" in Las Vegas, where casinos competed for customer attention with gigantic electric signs and displays. It only remained to complete the process of conflating nature and culture in Disneyland, where apparently real trees have plastic leaves, machines simulate alligators or African big game, and robots serve as actors in dramas. As Umberto Eco noted, for the visitor, "Disneyland tells us that technology can give us more reality than nature can."[51]

The power of electrical illusions is exploited across the West. Seen from an airplane or mountainside at night, its cities are glittering seas of lights, and each traveler encounters many postcard views of such landscapes, representing the city's glamour and mystery. In evening performances lights play on the geysers of Yellowstone (but not in the depths of the Grand Canyon). Thus electrification allows the westerner to exaggerate without words. Artificial light has become a characteristic part of the West's extravagant self-perception, mingling with its most potent natural landscapes and man-made symbols and eroding the sense of what is natural, possible, or real.

THREE
Domestic Landscape: Wright Morris's
The Home Place

The plains west of the Mississippi were first intensively farmed after the Civil War. The agricultural landscape Americans created in that space was a central part of the fiction of Willa Cather, Hamlin Garland, O. E. Rolvaag, and a host of other writers. In these works, as in Turner, the farmer was the central figure and the chief opposition to settlement came not from Native Americans but from the land itself, in plagues of grasshoppers, scorching sun, hard winters, tornadoes, drought, hail, and lightning. Those who survived built a rural society focused on small towns. They became central figures in a narrative of the transformation of American space into a distinctive landscape, in which farms were carved out of the land in neat quarter sections whose lines were laid out by the compass. The surveyor's geometry mapped a way of life based on isolated family farms, each imagined as a self-sufficient unit inhabited by a Jeffersonian yeoman farmer.[1]

The realities were far more complex. Farmers on the high plains found they could grow only a limited number of crops, and the further west they went, the less rain fell. In the late nineteenth century many believed that civilization and perhaps even precipitation followed the plow.[2] But by the 1890s angry Populists realized that their livelihood depended on railroad freight rates, ownership of grain elevators, bank interest rates, and international demand for agricultural products, as registered by the Chicago Grain Exchange. After 1900 rural farmers experienced a two-decade "golden era" with good prices and growing demand, but after 1920 came hard times, followed by drought and terrible times in the 1930s. The fundamental problem was that farmers produced too much, driving down prices. New technologies, such as tractors, trucks, harvesting equipment, and electrical machinery, made it possible for each farmer to raise more.[3] By 1948, when Wright Morris's photo-novel *The Home Place* appeared, rural America, which had been the center of the nation geographically, politically, and in more than one sense, spiritually, was rapidly being marginalized.

Nothing quite like it had been published by an American author.[4] Documentary collaborations between writers and photographers had become common in the previous decade, with some works such as *You Have Seen Their Faces* reaching a large audience. As William Stott demonstrates in *Documentary Expression and Thirties America*, this genre often exploited its subjects, reducing tenant farmers to types that illustrated economic statistics. The photographs in these popular works could strip people of any dignity, showing abject faces of defeat and ignorance amid a generalized landscape of superstition and poverty. The captions—such as, "Sometimes I tell my husband we couldn't be worse off if we tried."— purported to be the words of the people depicted but were often written by a middle-class, urban author, who was just passing through.[5] The genre both culminated and exploded in *Let Us Now Praise Famous Men*, where James Agee and Walker Evans treated share-cropper families as equals whose lives were as unfathomable and complex as their own. They evaded the easy generalizations and condescension of the documentary, but remained troubled about how to approach the lives of their subjects without violating their integrity.[6]

Morris eventually found his own solution to this problem by writing about and photographing the scene he knew best, his own Nebraska. Yet the autobiographical novel has its own pitfalls, and is by no means a guarantee of authenticity. Thus it is particularly interesting to know that several years before composing *The Home Place* Morris, like so many others, journeyed through the South with a camera. He gradually realized that the region was "a unique and a tormented culture . . . at once unavoidably visible and subject to instant falsification. The impoverished black, the debased poor white, had been well exposed in books and magazines, and such distinctions as might be made were in the eye of the beholder, not the camera." Morris grew more and more concerned about the validity of an image he or any other outsider might make in the South. Looking at a century-old home in Virginia, he realized, "The meaning this structure had to give out was a many-layered, many-voiced passage of history, too dense and complex to do more than acknowledge." When he made a photograph of that house, Morris "believed that what I had seen on the ground glass would surely be what I had captured." But as he took more photographs, he gradually realized that, "It would be weeks before I saw the negative, and many months would pass before I made a print of what I had seen on the ground glass. Would that image restore my original impressions, or would they be replaced by others? To what extent would this new image, cut off from its surroundings, constitute a new structure? How much of the 'reality' had it captured? How much had it ignored? Whether or not it had been my intent, I would end up with something other than what was there." He felt especially diffident about photographing people, and when he showed some of his early images to Roy Stryker, head of the Farm Security Administration's photography unit, "He listened with amusement to my explanation that the absence of the actual people enhanced their presence in the structures. ... He felt this to be a disturbing contradiction, the sort of thing you might get from a young writer."[7] Yet it was precisely this "contradiction" that would underlie the achievement of *The Home Place*.

Morris's first publication appeared in 1940, in an anthology which also contained an excerpt from Evans and Agee. Like them, Morris condemned the prevailing form of the documentary, and he would eventually seek another solution to the problem of linking text and image by writing a work of fiction in which he was both photographer and author, where he would write about his own relatives. Before this work appeared, however, some of these early images were published in his 1945 photographic essay *The Inhabitants*, which was still conceived in the documentary tradition. In such works photographs serve as illustrations, with captions clearly related to each image, as in a slide show with accompanying comments.[8] This achievement helped him to win a Guggenheim fellowship the following year, permitting him to return to Nebraska. There, with a new 4×5 view camera, he made most of the photographs which appeared in *The Home Place*, which, compared to *The Inhabitants*, was a more sophisticated modernist work, self-consciously Jamesian in its exploitation of point of view. The aesthetic which informs *The Home Place* was both a rejection of the 1930s social documentary and a challenging anticipation of postmodernist criticism.

The book can be briefly summarized as the account of a single day from just after breakfast until early evening in and around a farm. Its narrator, Clyde Muncy, has just returned home to Nebraska with his wife and two children, hoping to find a place to live and to write after being away for years on the east coast. During this day Clyde goes to the nearby town of Lone Tree, returns, and inspects the home of a dying uncle to see if his family could live there. Throughout the day he talks with relatives and savors the distinctive qualities of their lives. The book takes the form of dialogues, punctuated by his interior monologues. Clyde's consciousness frames the story, providing the illusion of direct contact with a rapidly disappearing agrarian world.

If one compares Morris to the American documentary authors and photographers of the Depression era, he emerges as a sensitive interpreter of rural life, fully aware of the prolonged farm crisis that still had not ended in the 1940s. But he refuses to foreground abstractions. Words such as "poverty," "drought," and "rural depopulation," never appear in his text. When Morris's narrator and his wife discuss economics with their daughter, they sound foolish to his Aunt Clara and the reader. Social problems are not ignored, but rather implied through telling visual details and apparently casual remarks. Because he grew up in Nebraska, Morris's writing shows a more intimate knowledge of farm life than John Steinbeck's *Grapes of Wrath*. The Joads are good people, but they are simple-minded and use more violent language than Morris's characters. He has a better ear for the crusty speech of the farm, with its dry humor, irony, and reticence.[9] Morris's farmers are complicated human beings, capable of irony and mystery. They are not merely survivors of the Dust Bowl who did not leave the farm for California. He treats their existence not as a lamentable social failure crying out for reform but as a complex local culture. Unlike city people they do not consume objects rapidly, but, if possible, use them for a lifetime, and to the narrator this is not anachronistic but worthwhile. The book evokes their rural sensibility, from the point of view of the one-time resident narrator who is sympathetic to the world he describes.

The rationale for this form of narration appears in the preface, a quotation from Henry James's *The American Scene*:

> To be at all critically, or as we have been fond of calling it, analytically minded—over and beyond an inherent love of the general many-colored picture of things—is to be subject to the superstition that objects and places, coherently grouped, disposed for human use and addressed to it, must have a sense of their own, a mystic giving out, that is, to the participant at once so interested and so detached as to be moved to a report of the matter.[10]

The novel which follows this preface at first seems far removed from Jamesian practice. Its 178 pages alternate between 90 black-and-white photographs and 88 pages of text, arranged so that neither is ever separate: a page of text always faces an image. Such a book is only possible using high quality semi-glossy paper, which does not allow print or image to bleed through from one side to the other. Print size and quality are also important, since small print would force a reader to pull a book too close to appreciate the photographs. With one significant exception, to which I will return, word and image are never mixed on the same page. There are no chapter divisions or formal interruptions of the action, and perhaps for this reason the *New York Times Book Review* compared it to a documentary film.[11] However, the work is not documentary, since its prose does not seek to inform us about social conditions in Nebraska, and its photographs are not presented as documents with some objective meaning but as subjective and partial views. Taken together, these images are not a story-telling sequence but a series of discreet visions: close-ups of personal possessions emphasizing textures; depictions of restricted interior spaces, showing the harmonious pattern of a few architectural details and furnishings; or views of farm wagons, outhouses, rural schools, windmills and other unmistakably rustic objects. Such photographs do not function as documentation and have no captions. They are only occasionally coordinated with the words facing them, and they are not given a fixed meaning or stabilized relation to the text. They cannot be understood as illustrations, and most of them fill the entire page to its edges, so that they are not framed by a white border, a heavy line, or any other form of visual closure. A frame around an image announces its completeness. In the conventions of art, a white border says that the picture is a self-contained statement. Morris's un-named and un-framed images fill the book with glimpses that cannot be illustrations, stills from a film, or autonomous works of art. By releasing these photographs from definition through language and from closure through framing, he makes them problematic and indeterminate.

A. D. Coleman has said of these images that "they are only infrequently tied to specific events in the narrative. Instead, they are used to epiphanize, by directly visual means, those subtly changing moods at which the narrator hints only obliquely. The initial result of this startling device is the gradual erasing of the thin, firm line between art and life."[12] To erase this line Morris developed a new version of the first person narrator, who is also a "first person seer." For James, point-of-view narration was not just a literary technique, but a method that was realistic because it was true to the human experience of subjectivity.[13] Morris's narrator adds a new dimension to this subjectivity by supplementing language with photographs taken from the stance and perspective of an adult of normal height walking through the world of the text. There are no images which could not be the result of ordinary vision—no wide-angle shots, no collages, no aerial views, no solarization, no obvious tricks. His camera works like an eye and its images have the feel of remembered objects and landscapes, fusing image and narrative consciousness in a transcription of living in the world.

In the first line of *The Home Place* the narrator asks—"What is the old man doing?"—because his eyes are poor. His aunt, who has a glass eye, guesses that the old man may be planting melons. His son can see well enough to declare that the old man is "fixing his inner tube." Judging by the photograph facing page one, the boy is right. His great uncle stands before the barn door holding a patched inner tube in his hands. In a book that will focus on the social life of objects, this first example is a lesson about the inadequacy of either vision or narration alone. Subsequent dialogue discloses that the old man is not fixing the inner tube, but only condemning it as an irreparable modern object made of "airsuds." He sees the object in the light of experience. Shortly thereafter, Morris refines the point that good eyesight is not equipment enough. When the boy finds a croquet ball, he doesn't know what it is or how to talk about it. If this work begins by showing that the adult narrator sees poorly and that photographs are insufficient evidence of what people are thinking or doing, it also quickly eliminates the romantic child as seer.

In this opening sequence Morris poses two central problems. The first is the relation of photograph to text, of what an image might best depict when it forms part of a work of fiction. Some subjects are not visualized. In pointed contrast to all documentary photographers, the book never presents a human face, and in only four images does a figure appear at all, in each case the old man. In the first we see him frontally, his face hidden entirely in shadow, as he bends over the inner tube. The second shows him from behind in strong sunlight against a dark background, wearing an old black hat. The third shows him seated in the doorway to the barn, his face completely hidden in the shadow of a white straw hat. In the book's final photograph, the old man is taken in the same doorway as in the first image from a step further back. Here he walks into the shadows of the barn, right leg already disappearing within the door. The first, third and final images contain an interior framing in the form of the barn door, while the second highlights the old man against a black background. This repeated refusal to show the face testifies to Morris's respect for his subjects, and his refusal to subject them to the prying finality of the camera. Indeed, early in the book he pointedly criticizes urban people for their "vulgarity" in contrast to country people who "take a man at his face value, as they figure it's his own face, a fairly private affair, and the only one he has. They don't roll the eyelids back to peer inside of it. They don't leer at you with the candid camera eye." In hiding the face, the subject becomes certain textures: smooth old rubber, rough woods, trampled straw, well-worn clothing that takes the mold of his body, distinctive hats. Such images shift attention from the old man's appearance to the textured objects of his world.[14]

The aesthetic of these images is thus directly related to the second question Morris poses in the opening "tire" sequence: what is the relationship between people and their possessions? Virtually every object depicted has a known use, but this functionality is not its meaning. For example, Aunt Clara pickles beets, standing over the stove, and only after this is established in the dialogue do we see it. But the photograph does not show the stove in use, with the mason jars spread around on the counter and a large kettle smoking on the burner. That sort of image would be literal, banal, imaginable. Morris shows the stove worn with use.[15] It is not just a place where Aunt Clara cans beets, but the place where she has cooked for fifty years. He presents the stove not merely as a setting or a background comprehensible in terms of utility. It is a central site in a world of household objects with a collective pattern and meaning that grows out of repetitive human use. It is also a possession that Aunt Clara would prefer visitors to see only when tidy. Just as Walker Evans was scrupulous about letting subjects dress and prepare themselves before he photographed them, Morris never exploits a temporary sloppiness to create an image of squalor or untidiness. The stove, scrubbed and shining despite years of use, the ash bucket empty beside it, looks as it would if Aunt Clara had visitors coming.

Many of the images show objects so common as to need no naming and yet so uniquely part of the domestic landscape as to be almost beyond the powers of descriptive language. Agee was well aware of this problem, and tried to solve it through an exhaustive catalogue of the minutiae in a sharecropper's home, respectfully listing the objects he found in each room, including the drawers. Morris chose synecdoche, letting a few photographed objects stand for the whole. Not all intruders were so sensitive. Bourke-White recalled that once she and Erskine Caldwell "went into a cabin to photograph a Negro woman. . . She had a bureau made of a wooden box with a curtain tacked to it and lots of little homemade things. I rearranged everything. After we left, Erskine spoke to me about it. How neat her bureau had been. How she must have valued all her little possessions and how she had them tidily arranged her way, which was not my way. . . . I felt I had done violence."[16] Morris, like Caldwell, knew the rural world, and took pains not to disturb but rather to celebrate the harmonies its inhabitants had constructed. He photographed knives and forks spread on a newspaper in a kitchen drawer; the arrangement of personal items in a bedroom; a man's clothing worn with use hanging on pegs; clippings from newspapers pinned to a wall; a shaving mug and two brushes resting on a white towel.

The ineffable particularity of the domestic landscape plays an active part in a central scene of *The Home Place* in which the narrator and his wife, who need to find a place to live, realize they cannot take possession of a house. They enter the vacant home of their dying relative, a man whom they hardly know and who is in a nursing home. On coming in, they realize that the house is not really vacant.

> . . . any house that's been lived in, any room that's been slept in, is not vacant any more. From that point on it's forever occupied. With the people in the house you tend to forget that, the rooms and the chairs seem normal enough, and you're not upset by the idea of a FOR RENT sign. But with the people gone, you know the place is inhabited. There's something in the rooms, in the air, that raising the windows won't let out, and something in the yard that you can't rake out of the grass. The closets are full of clothes you can't air out. There's a pattern on the walls, where the calendar's hung, and the tipped square of a missing picture is a lidded eye on something private, something better not seen. There's a path worn into the carpet, between the bed and the door, the stove and the table, and where the heel drags, the carpet is gone, worn into the floor. The pattern doesn't come with the house, nor the blueprints with the rug. The figure in the carpet is what you have when the people have lived there, died there, and when evicted, refused to leave.[17]

In continual reference to James, in other passages Morris plays with the idea of a "figure in the carpet" that wears out. While talking to Aunt Clara, he notices that "on the floor was a piece of worn linoleum. The center of the pattern had been worn off, and Clara had daubed on one of her own. Brown and green daubs of the brush." In the final lines of the book, the narrator describes "a white cotton glove, with the fingers spread, thrust up in the air like the gloved hand of a traffic cop. The leather palm was gone, worn away, but the crabbed fingers were spread and the reinforced stitching, the bib pattern, was still there. The figure on the front of the carpet had worn through to the back."[18] The recurrence of such figures in these artifacts underlies the book's aesthetic. Human personality imprints itself on the visible world through the use and arrangement of possessions. The narrator insists that: "The carpet wears out, but the life of the carpet, the Figure, wears in. The holy thing, that is, comes naturally. Under the carpet, out here, is the floor."[19]

These passages show how Clyde as the interested yet detached observer captures the "mystic meaning" or "the holy thing" in the objects of human use at *The Home Place*. Understanding this purpose clarifies the relationship between Morris's photographs and his text: a relationship of double framing. The language evokes the life flowing around these objects and shaping them to human use. In this sense, language serves as a framework surrounding and shaping the photographs' meaning. But the photographs also frame and record the life of the farm. They fix and hold a pattern of habitual actions, recording the accumulated experience of the people who live there. Thus the combined frameworks of language and photography do not refer to each other, but each independently refers to the old homestead, together framing it both in the flow of time (language) and in the outlines of space (photographs). The numinous talismans of this world are everyday objects worn with use that can be only partially depicted and only partially explained. Their existence is almost beyond reach, but retrievable, as text and image implicitly offer the reader/viewer the chance to infer the reality of *The Home Place* by triangulation. The relationship between photograph and text is not one in which the words interpret the meaning of images, nor is it one where images illustrate the prose. Instead, Morris uses them to supplement each other, as he stated in his first photo text from 1940: "Two separate mediums are employed for two distinct views. Only when refocused in the mind's eye will the third view result. The burden of technique is the reader's alone."[20] Together, text and image affirm the world's recoverability in art.

In *The Home Place*, however, Clyde guides the reader to the appropriate response as he is captivated and moved by the everyday objects of his relatives. His sensibility can be traced back to both the Transcendentalists and English Romantics. In his Uncle Ed's home he knows that "There was not a thing of beauty, a man-made loveliness anywhere. . . " Yet seeing the house struck him the way "I expect a thing of beauty to make me feel. To take me out of myself, into the selves of other things."[21] The clearest possible example of this unexpected beauty in the ordinary, and of the relationship of world, image, and text occurs when Clyde and his wife enter Ed's bedroom, and "she stepped back, as if she saw someone in the bed."

There are hotel beds that give you the feeling of a negative exposed several thousand times, with the blurred image of every human being that had slept in them. Then there are beds with a single image, over-exposed. There's an etched clarity about them, like a clean daguerreotype, and you know in your heart that was how the man really looked. There's a question in your mind if any other man, any other human being, could lie in that bed and belong in it. One might as well try to wear the old man's clothes. His shoes, for instance, that had become so much a part of his feet they were like those casts of babies' shoes in department stores.[22]

The camera's privileged relationship to the objects of this world is now clear. In Morris's aesthetics, a bed can daguerreotype the body, revealing its essence. What is true for the bed, the clothing, and the shoes, is true for the house as a whole. It is a vast negative with a very slow exposure time that has taken the imprint of a man's life. Exposing the camera to this bed transfers the "negative" to a latent image on film that can be developed into a photograph, preserving Uncle Ed's shape in bronze. The photographer in such a system must cultivate a new form of the romantic sensibility that quite literally requires "negative capability," since he does not create but only apprehends the latent image lurking in the object and then captures it on film. The linkage between the object, the camera, and the image is thus transformed from an act of interpretation into a replication of subjective truth, whose validity is ensured by the camera's supposed mechanical veracity.[23]

The photograph of Uncle Ed's bed which accompanies this passage thus becomes far more than a verification of its physical existence. It reveals the impress of his body on the mattress, suggesting his weight in the permanent sway of the springs. Through use he has transformed the bed into an extension of his being-in-the-world. Clyde and his wife refuse to sleep in this bed or to move into his house, because his presence is too strong. In the same way, Morris expects that the reader cannot entirely enter the world of these rural farmers. Their objects finally resist assimilation to the general categories of "bed," "shoes," or "vacant house," and must be treated as unique and separate things. They cannot be framed and turned into aesthetically complete art objects. That is why the first and the last images foreground clothes rather than the man. Morris does not mean to say that "clothes make the man," but rather that the man's living makes the clothes uniquely his.

But note that not all of reality is stenciled with the human personality. Not every arrangement of objects contains the impress of a lifetime of self-expression. Throughout *The Home Place* Morris evokes the difference between the urban world that the narrator has left and this rural world of human solidities. Just as there are motel beds where the image of living has become hopelessly blurred by a succession of bodies, so the city consistently functions as an alien landscape where objects have become harder, standardized, and synthetic. Indeed, the purpose of this return to an American arcadia is to give the children a better place to grow up in, a "more real" place.[24]

Both in its language and its photography *The Home Place* is superior to the documentary tradition of the 1930s, and on a par with *Let Us Now Praise Famous Men*. Yet it failed to initiate a new genre, and this failure is intimately connected to its aesthetics. The usual explanation for this failure—the lack of individuals equally gifted as novelists and as photographers—is suspect, since Morris himself did not work in the genre again, but instead wrote twenty volumes of novels, short stories and literary criticism. The reason his particular form of the photo-novel was still-born lies in its conservative version of modernist aesthetics.

Here a European perspective is helpful. Within the American context, Morris can be seen as the inheritor of Walker Evans and the best of the Farm Security Administration documentary photographs. But from a European perspective, he bears a curious relation to the photographic theory of defamiliarization current in the 1930s, which assumed that photography could teach one to see the world afresh by "making strange" an everyday reality encrusted with banality. The underlying notion behind this aesthetic, as Simon Watney has pointed out, was that the organs of perception were "all-conditioning portals of knowledge" that could be "cleansed of the mystifying and misleading accretions which result from our experience of a corrupt and corrupting world."[25] Alexander Rodchenko and other practitioners of this influential school "made strange" their subjects through the use of odd angles, mirrors, extreme close-ups, distorted shadows, juxtapositions, and collage, a virtual catalog of all the techniques Morris eschewed, with the significant exception of the close-up, the only one of these practices that is also compatible with his adaptation of Jamesian realism. The purpose of the many close-ups in *The Home Place* is not so different from that of the defamiliarization school. Morris forces our attention on three horseshoes, the center of an old wagon wheel, an ear of corn stuck through a ring on the wall, a patch of cracked mud and dry grass, the steering wheel of a car, a basket of corn-cobs, a tombstone with no name, a stack of worn-out rubber tires, and more. He grabs the reader by the neck and says, "Look at these," without giving the slightest reference to them in the text. None needs to be named, but by implication all can be seen anew. The approach might be termed "re-familiarization."

But if one accepts Morris's aesthetic on its own terms, there is an irony in his particular selection of objects to valorize. The material world *The Home Place* idealizes is not the result of a society producing hand-made objects, but rather one which bought flatware, furniture and mason jars from the Montgomery Ward catalog.[26] Its stoves, carpets and windmills were produced in bulk and sold ready-made. Morris relies upon a reader's nostalgia for a pre-lapsarian universe of object relations, organized according to an agrarian system of values only partially visible to the outsider. One's very distance from this world permits the invention of an elaborate fiction about it, a refurbished version of the pastoral myth of a past when people lived in the landscape. The objects of their world are not presented as consumer goods like the tire made of "airsuds" which the old man criticizes in the opening sequence. Rather, these objects last a life-time, constituting a world where men and women put their own figure into the carpet, as they wear through it down to the floor.

Morris employs both text and image to undermine the more exploita-tive forms of 1930s documentary photography, but he does so in order to recover an unalienated world where family values are embedded in the artifacts of everyday life. His book is a kind of hyper-realism, expanding the single narrator and Jamesian point of view to include a subjective photographic version of the world, in which the romantic ideal of the poet's "negative capability" makes a stand against the steady erosion of fixed meanings. But in this new version of Coleridge's aesthetics the machine itself, in the form of the camera, must assist language to recover a dying portion of the world.

Two decades later, in *God's Country and My People*, Morris returned to these same photographs, which are now all that remains of the farmstead, to meditate upon them.[27] In this third and final reworking of the images, they awaken his memories, evoking a prose of fragmentary reconstruc-tions. In accordance with the passage of time, his romantic aesthetics of presence ineluctably leads to a disappearing subject. The equilibrium between word and image in *The Home Place* breaks down as text becomes subservient to image, and neither refers to a present world. The photo-novel thus decomposes into a story that cannot be continued but only rewritten. This result confirms the post-modernist theories of Roland Barthes and Jacques Derrida about the inadequacies of realism, even when photography supplements language.[28]

Yet already in 1948 Morris knew how photographs lose their specificity and meaning over time. Near the end of *The Home Place* text and image appear together for the only time, and the characters make direct reference to this image, which in fact was taken of the Morris family near Zanesville, Ohio, c. 1895. In the novel this photograph is from 1892 and it presents a lesson in the historicity of photographs. The old photograph tacked to a rough paneled wall shows fourteen well-dressed people standing rather formally in front of a white frame house with a picket fence. Its meaning resides not in any formal aesthetic qualities, but rather in its representation of former family members to those who survive. The oldest remaining, an old lady who has come to visit, demands, "Where's that picture?" And to the reply "What picture's that?" she snaps back, "'Now you lookee here,' she said, 'you know very well there's just one picture.'"[29]

"Just one picture" from 1892 occasions ten pages of dialogue, mostly about the people depicted, with little said about the photograph as a photograph. Even that is turned into a comment on the family.

> "Lord—it's fadin'!" the old man said.
>
> "You wearin' your glasses?" Clara said, and covered her eye to look at him. He had pushed them back on his head. He let them down on his nose.
>
> "Come to think —" he said, "they never had much as faces. Been thirty years since they had faces." He moved his glasses up and down his nose, focusing. "No, it's worse!" he said.
>
> "Most of us dead and gone, think it would be fadin'!" Clara said, "Same as me an' you are fadin'."[30]

If the old man can barely see and fades with the image, in his mind's eye he can reconstruct each person and tell an anecdote about them. For him, the picture works in the same way as the other images of *The Home Place* are meant to function for the reader. They evoke memory, and together with language permit the recovery of the past by an interior triangulation This aesthetics based on human presence would seem to be at the farthest remove from theories of deconstruction. Yet long before Barthes or Derrida, Morris knew that any reconstruction of the intimacy of an earlier time was a fiction that could fade and die with its makers.

II

NARRATIVES

FOUR
*Reconstructing Rural Space: Four Narratives
of New Deal Electrification*

The landscape of Wright Morris's *Home Place* was rapidly being transformed even as he wrote in the 1940s. The horses were mostly gone, replaced by tractors and trucks, and electricity was coming in as fast as farmers could get it. From the 1930s on several competing narratives described the New Deal rural electrification programs that played a central role in the reconstruction of rural space. One can tell this story from at least four points of view: the social critics of the 1930s who called for a revitalized countryside; the private utilities; the government agencies, notably the Rural Electrification Administration (REA) and Tennessee Valley Authority (TVA); and the farmers who received electrical service for the first time. (I will provide occasional examples from the Indiana countryside surrounding Muncie, Indiana, site of the many Middletown studies.[1]) All of these narratives begin with an unelectrified rural world and end with a region that has been wired up to the rest of the nation. So these are all stories about integration or homogenization, but there the similarity ends.

I
The narrative about the New Deal electrification that perhaps most easily comes to mind half a century later is an ironic tale. In it, carefully planned actions have the opposite of their intended effect, as government attempts to reverse a long-term trend away from farming fail miserably. One of the most striking changes in the last two centuries of American history is the shift away from agricultural life. In 1800 roughly 95 percent of all Americans were farmers; today less than two percent are. The scale and

organization of individual farms has been changing continually throughout these years. Productivity per farm had to increase tremendously for such a change to take place. In this long perspective, rural electrification exemplifies the substitution of capital equipment for labor, bringing both reductions in waste (for example through refrigeration) and improvements in productivity. The irony enfolded within this success is, of course, that rural electrification was not implemented to drive people off the land but to keep them there, by bringing to the country the amenities enjoyed by the city. Yet paradoxically, rural electrification meant fewer farmers were needed. We thus have an ironic tale, in which the historical actors undermine their goals by their actions, so that the apparent, immediate meaning of events (a rural revival) is eventually undercut and contradicted by long-term results.[2]

Such a story must be told as part of two time frames, one close to the events, the other stretching over half a century or more. The tale within the larger frame could well begin in 1935, when the REA began its work. Many felt that American cities were overcrowded, the places of the proverbial apple sellers, breadlines, and unemployed. As Paul Conkin put it, "The powerful urban magnet, which had drawn millions of farmers into its enticing fold during the twenties, had lost much of its attraction. As progress slowed, many men turned their eyes back to the land, to the old homestead, to security, to a memory."[3] In Muncie the Depression hit especially hard. Almost one-fourth of the manufacturing establishments closed. By 1933, the businesses that remained paid only one-third as much in wages as they had in 1929. Many people returned to the land, and in the early Depression years the number of farms in Muncie's Delaware County actually grew by nine percent, despite an increase in foreclosures.[4] In the midst of the Depression the movement from the country to the city apparently had reached its limit. A historic reversal of population movements seemed imminent.

As the Depression deepened, a growing number of social critics reenforced this trend. They advocated getting "back to the land," and interest grew in figures such as Ralph Borsodi. His two influential books from the 1920s, *This Ugly Civilization* and *Flight from the City*, encouraged people to take up subsistence farming. Brosodi had done so in 1920 and made a success of it.[5] Another sign of the reaction against urbanization was the emergence of the Southern Agrarians and their call for a return to rural and regional values. John Crowe Ransom, for example, "wanted to decentralize the service of agriculture and move back, not only toward more subsistence farming, but to stronger, less dependent regional marketing systems, to the type of rural economy that prevailed in the late nineteenth century." He realized that electricity could make farming more attractive and advocated rural electrification.[6] His fellow Agrarian, Herbert Agar, hoped to combine urban amenities such as electricity with small-scale factories and subsistence farming. Frank Lloyd Wright developed the idea of "the new city that will be nowhere, yet everywhere." His "Broadacre City" was a decentralized form made possible by electricity and the automobile. Likewise, the National Catholic Rural Life Conference advocated "the replacement of steam power with electricity." This "would make feasible the development of small factories and home-style industries and, by facilitating communication, would also tend to integrate the rural areas with national life."[7] Henry Ford made similar arguments and even built a few small factories in the Michigan countryside.

Nor were these ideas limited to critics and outsiders, for many urban planners who influenced the New Deal shared an antipathy to the city. Lewis Mumford, Clarence Stein, and Louis Wirth called for green belts, satellite cities, and re-creation of a community environment. Mumford wrote the screenplay for a film, *The City*, which contrasted living conditions near heavy industrial factories with an idyllic new green-belt community where cars were minimized and homes were near both school and work. Rexford Tugwell, director of the Resettlement Administration, believed the nation needed several thousand such planned cities, although only three green-belt towns were built. However, the Federal Emergency Relief Administration moved many farmers from marginal land and helped them to relocate on more fertile soil, and later the Farm Security Administration built 154 resettlement projects and also loaned money to farmers.[8]

Yet while people as diverse as Borsodi, Ford, Wright, Catholic reformers, Southern Agrarians, and New Dealers agreed on the desirability of rural electrification as a means to regenerating life in the countryside, relocation programs and rural electrification instead helped make farmers so productive that many became redundant. While 40 percent of the population was on the farm in 1930, only 15 percent remained in 1950, and 8.7 percent by 1960. By the 1950s the REA suffered from a declining number of customers and new government programs were started to attract industries and businesses into the emptying countryside.[9]

This ironic reading of rural electrification may be premature, however. The long-term effects of creating a national grid only became fully apparent in the census figures of 1970 and beyond, when it became clear that Americans were not merely spilling out of cities into suburbs:

they were moving ever further out, gradually erasing clear lines between city and country. Indeed, population density in American cities peaked in the early Depression and has been steadily declining since then. Studies of Denver, Milwaukee, and many other cities show that at the same time as people were moving off the farms there was also a net migration out of downtown. This process was already underway before World War II, more than halving the density of urban settlement by 1963.[10] By 1975 migration into rural America had become greater than out-migration, and the trend shows no sign of abating.[11] According to a 1985 Gallup Poll, more than 60 percent of all Americans now want to live in small towns or in the countryside, and many are acting on that desire. The *New York Times* reported in 1988 that more than 500 rural counties were experiencing 14 percent growth in the 1980s alone, due to the in-migration of wealthy families from the suburbs.[12] Americans have opted for a modernized middle landscape, between farm and city. The diaspora has been further assisted by the advent of home computing, fax machines, and modems. Without rural power none of these machines would be useful beyond the edge of suburbia. Because the rural cooperatives own 45 percent of all American electrical distribution lines, at their centenary in 2035 they may supply a large, decentralized service sector and high-tech light industry. If the result is not precisely Wright's "Broadacre City" or the regional revival envisioned by Southern Agrarians, the city has not triumphed over the countryside. Rather, urban deconcentration has triumphed over both. The political loyalties of these ex-suburbanites are not yet clear, but there seems good reason to believe that they will not be populists or liberals supporting the reform wing of the Democratic Party, but rather cultural conservatives in flight from social problems. This was hardly the constituency Franklin D. Roosevelt set out to serve.

II

Looked at from the point of view of private enterprise, rural electrification is another story, one of rational free-market development frustrated by government interference. This story has not been popular with historians, but private utilities and some members of the Republican Party cling to it. Its presuppositions are formalized in Harvard Business School textbooks. To utility executives before 1930, rural customers were a low density, unprofitable market. However, the electrical interests were not monolithic. By the end of World War I at the latest, General Electric, Westinghouse, and many electrical supply companies aggressively marketed a complete line of goods for the farmer. Farms seemed to offer an enormous untapped market for electrical motors and appliances. For example, in 1916 a company in Cleveland, Ohio, issued a 128-page electrical catalog aimed at customers beyond utility lines.[13] But while equipment manufacturers and hardware stores advertised their wares to farmers in national magazines and displayed equipment and appliances at county fairs in the 1920s, private utilities used most of their resources to electrify urban homes, while sponsoring only a few experimental rural projects. The long distances between farms, with only three or four customers per mile, made transmission costs high, and farmers had little money, due to the long agricultural depression that persisted throughout the 1920s. Urban markets, in contrast, were packed with potential customers who had discretionary income. In 1910 only one in ten urban homes had electricity, and utilities maximized their profits by reaching the other 90 percent of them by the end of the 1920s. When the Depression began, private power companies were moving slowly into rural areas. Creation of regional grids between the major cities facilitated this process somewhat, as did economies of scale in power generation.

The Depression had far less effect on utilities than is generally understood, as domestic electrical consumption soared during the 1930s. Utilities benefitted from the structural realignment in the economy, which had already begun in the 1920s, as heavy industry gave way to services and high-tech areas such as airplanes.[14] (Both trends were conspicuous in Muncie, a city primarily based on heavy industry.[15]) From the business point of view, electricity was not an instrument of social policy and agricultural reform; it was a commodity to be produced and distributed for profit. Both this conception of electricity and private ownership of power itself came under attack in the 1930s. In Muncie a socialist was elected to the city council and together with a reform-minded mayor and local citizens groups agitated for a return to municipal power. Petitioning citizens forced the issue to be placed on the ballot for referendum, but the local utility successfully petitioned the courts to enjoin the election, on technical legal grounds. This local conflict suggests rather well the public mood in the early 1930s, when the affluence of private utilities contrasted sharply with an empty public treasury, and when many called for state ownership of public services. In this struggle the general lack of rural electrification in the United States in contrast to other industrial nations made the utilities particularly vulnerable. Roosevelt's election gave utility reformers a chance to tap the public mood and develop new forms of public power.

In Muncie this resulted in a race between newly formed cooperatives and the local power company, as each tried to sign up as many farmers as possible. The utility built more lines into rural areas and reduced its service rates, so that after April 1936 rural and urban customers paid the same. Nevertheless, new rural customers had to guarantee that each year for five years they would consume electricity worth 18 percent of the cost of line construction. This provision was designed to make sure that the utility's initial investment was recouped within six years, and it kept smaller customers away, while taking from cooperatives some of the most lucrative possibilities.[16] Indeed, cooperatives protested the construction of these "spite lines," charging that by skimming off the cream of the customers and making in-roads in rural areas, private utilities prevented rational market development. The utilities saw the matter differently. They were pursuing a fifty-year-old policy: reaching the most profitable, large customers first, to be followed by later line extensions to smaller customers. For utilities, rural poverty was not an incitement to action, but a caution against rapid expansion. New Deal programs seemed to rob them of future markets, by granting low interest loans to cooperatives and selling them inexpensive electricity from publicly owned dams. In this conservative view, the federal government had violated free development of the market, artificially sustaining marginal farmers, who in turn over-produced, thereby requiring yet more federal assistance. In its most dramatic form, this story posed the forces of Americanism and capitalism against demagogic European socialism. *The Muncie Morning Star* editorialized in this vein: "Congress must decide soon whether American principles are to protect industry or the country is to be blighted with the New Deal's brand of state socialism. . . .The [TVA] program . . . should be rejected by an overwhelming vote if private initiative is to be salvaged from the wreckage of impending socialism."[17]

III

Of course there is a counter-myth to this Republican story, in which the Democrats, lead by Franklin D. Roosevelt and his men, notably Morris Cooke, John Carmody, and Henry Slattery, battled the private utilities. New Deal agencies brought modernity to rural people across the land, improving health and sanitation, dispensing loans that helped the local economy, building cooperatives that stimulated grass-roots democracy, and creating a yardstick to measure the performance of private power companies. The REA and the TVA, as vividly portrayed by the photographs of the Farm Security Administration, both literally and figu-ratively carried light and prosperity to suffering rural people. In this epic the heroes are the linesmen, the farmers who organize, and above them all, the far-sighted politicians in Washington. Proof of the virtue of the enter-prise was the fact that most of the loans, including those in the Muncie area, were eventually paid off. The electrical cooperatives were financially successful. This has become the standard version of the meaning of the REA, continually retold by the National Rural Electrification Cooperative Association and echoed in much of the literature, notably in Marquis Childs's *The Farmer Takes a Hand* and D. Clayton Brown's *Electricity for Rural America*.[18]

This tale of rural electrification foregrounds the TVA and REA as central parts of the political struggle to determine the role of the state in the economy. This story has often been told, and here I can offer only a brief summary. The REA and TVA were hardly identical.[19] The TVA generated power at large dams, and sold its own power to both private and public customers, including farmer cooperatives. The REA initially had few generating facilities and primarily built distribution lines served by existing power companies.[20] The TVA was focused in one area of the United States; REA was spread over most of the nation, and consisted of many small cooperatives.[21] The administration of the two agencies was also quite different. The TVA began with utopian ideas, in part inspired by Edward Bellamy, about transforming a poor region into an industrial-agrarian utopia. The cooperative form of organization was adopted in part to further ideological goals of constructing a new society.[22] In contrast, the REA began by trying to work with the large power companies. When negotiations broke down, the agency then adopted the cooperative form of organization less for ideological than for practical reasons.[23]

Nevertheless, the two agencies converged over time, as the TVA's idealism was blunted by changes in administration and the REA was mildly infused with a blend of populism and socialism. In the National Archives, for example, are speeches given by REA officials from the later 1930s praising Scandinavian social democracy and its cooperative movement.[24] But as a practical matter, both agencies had to put their business on a firm financial footing, and to do this, they soon discovered, meant working closely with appliance manufacturers and local businesses. Signing farmers up in cooperatives and getting electric lines to their homes were only first steps. To make a cooperative self-sustaining required increases in electrical consumption, and REA coops soon were being visited by "utilization experts," Washington's bureaucratic term for salesmen. Their mission was to convince farmers to buy motor-driven equipment and larger appliances. Like the utilities before them, cooperatives soon realized that farmers who never installed more than a few electric lights cost more than they were worth. Coops had to sell more power or they would never be able to pay back their government loans. They were not part of a state-run economy but small units within the structures of capitalism. Cooperatives had been created by government intervention in the marketplace, through inexpensive loans and technical assistance, but their fate clearly depended on achieving a balance between sales and expenses. While some money could be saved in line construction by asking that farmers donate their right of way, and while some poor customers could be brought into the grid by putting them to work on line construction, these initial savings had to be followed up by growth in electrical consumption.

Here, Washington agencies at first misjudged the farmers, as the "utilization experts" initially pushed the sale of heavy farm equipment. In practice, however, only a few kinds of farms, notably dairies and chicken ranches, discovered immediate advantages in electrical equipment. Raising wheat, tobacco, cotton, and many other crops could go on profitably much as before. But the farmhouse was quite a different story. Farmers adopted washing machines and refrigerators more quickly than urban consumers, and they did so for sound economic and practical reasons. In the city laundries competed with home washing machines; in the country all washing had to be done by hand after water was pumped from a well and heated on the stove. Little wonder half of the REA's customers in the first three years bought washing machines, which had a much higher priority than indoor toilets. In Henry County, just outside Muncie, only 11 percent of the newly electrified farms had flush toilets, but two-thirds had washing machines, and 40 percent bought refrigerators. A 1939 study of Boone County, Indiana, the first cooperative in the State, showed that after two years of service 68 percent had washing machines while only 10 percent had utility motors; 45 percent had vacuum cleaners while less than 1 percent had bought any machines for agricultural production.[25]

Where urban consumers could go to the store every day and have ice delivered as needed, most farmers had no service, and many lived in a climate where no ice could be harvested. A refrigerator could save produce from spoiling and eliminate the need for some trips to town.[26] In the late 1930s REA officials found that the average cost of owning a refrigerator had dropped since 1928 by almost 50 percent to $165 and that the machines lasted eight years, with low upkeep costs. The annual cost of ownership was thus less than the $25–30 per year it cost for electricity to operate it.[27] As such cost accounting suggests, the REA cooperatives soon regarded farmers less as members than as consumers. In the early 1940s the REA drew up a model plan for electrifying a dairy farm ($2000) that included not only wiring the house and barn ($500) but a complete line of home appliances ($630) that cost more and used more energy than the dairy ($525) itself. The dairy farmer also was encouraged to invest in quite possibly unnecessary technologies, such as electric pumps for garden irrigation ($40), an electric brooder ($30), an electric fly trap ($15) and an "Electric Hot-Bed" ($40). Not incidentally, each of these heating devices used a great deal of electric current.[28] The farmer could borrow $2000 for these appliances from the government at a rate of 2 percent, but he could not borrow additional money to pay his suddenly increased utility bill. Instead, he had to increase production. To the extent that a fully electrified farm was more competitive, it drove less efficient operations out of business.

The marketplace that had governed private utility patterns of investment ultimately controlled the cooperatives as well. Electricity remained a commodity, whose sale the government had extended into new areas. Doubtless the individual farmer was better off, and unquestionably electricity had come more quickly than it would have through purely private initiative. But in a long-term perspective, public rural electrification converged with private electrification. In later years cooperatives increasingly relied upon full-time managers and increasingly they promoted electrical sales as an end in themselves. Where once the rural people came to meetings with a sense of pride, proprietorship, and participation, two generations later most members regard coops as local utilities. The demands of the market gradually subverted the idealist hopes of stimulating grass-roots democracy or building an electrified rural alternative to the city.

IV

"All we had was wires hanging down from the ceiling in every room, with bare bulbs on the end. Dad turned on the one in the kitchen first, and he just stood there, holding on to the pull chain. He said to me, 'Carl, come here and hang on to this so I can turn on the light in the sitting room.'"[29]

As in this anecdote, farmers were sometimes represented as hicks by REA publicity, which carried stories of women who did not know the lights could be turned off and of men who were afraid to change light bulbs. While no doubt a few thought electricity would drip out of their sockets like water, most rural people knew a good deal about it. The telegraph was quite familiar after a century of use, and it contained the basic principles of later electrical installations: a current that ran instantaneously through wires over long distances, creating an immediate connection between two points. Furthermore, most rural people had relatives or friends in towns, where electricity had become almost universal. While the REA liked to send out colorful publicity about rural naiveté, many farmers, with some technical assistance, built distribution lines, and often they were able to wire their own homes. These amusing REA stories catered to urban fantasies about simple people in the countryside. They were part of the Jeffersonian myth of self-sufficient farmers who had little contact with the industrial world.

Such anecdotes emphasized rural poverty and ignorance, which it was also possible to illustrate through photographs. Many images were designed to show how federal programs were redeeming both city and farm. Only a few of the FSA photographers, notably Walker Evans, took an interest in the vernacular and traditional elements of local life that were being lost. While Evans himself was a middle-class outsider, he made a point of photographing the hand-made, the traditional, and elements of folk art. He sought out the indisputably local: the painting of a bull on a brick wall, a tattered minstrel showbill, a stamped tin relic, or a wooden church made from rough boards. His photographs often visualize the meeting point between a local culture and industrial civilization, such as an African-American barber shop, whose walls are covered with newspapers, or the interior of a general store where local and mass-market goods intermingle.[30] Evans's perception of the split between folk culture and industrial civilization complicated his attitude toward technology. It also qualifies the triumphalism of government agency publicity. Evans's images testify to the rural world we have lost, and hints at another story of rural electrification, in which a way of life vanished, as irrevocably as the waters of a dam covered some communities.

Paul Conkin edited a volume that attacked the TVA for being a "grass-roots bureaucracy" which imposed new conditions from above, despite its rhetorical commitment to the common man. The resistance of old settlers to TVA dams was not avidly recorded by the New Deal, but traces of it can be found in a variety of documents, ranging from oral histories and local newspapers to some fiction. For example, Eleanor Buckles' novel, *Valley of Power*, contains powerful descriptions of the unimproved rural landscape which local residents do not want to leave. It may be a hardscrabble life, but they are comfortable and independent. They know how to make a living, farming, cutting down trees and trimming them to be railroad ties, and they have rituals, such as Saturday night dancing, coon hunting in the woods, and fishing, which link them into a community. They are rooted in the place. As one father explains to his daughter, Sandrey: "Look out across the hills and hollows and see how good it is; listen to the stirrings of live things in the grass, and smell the good smell of it. I never wanted this just for me. I wanted it because my Pa's Pa gave it to him, and my Pa gave it to me, and I wanted to hand it on to my young ones. . . But . . . you're reaching out somewhere beyond the place you came from—somewhere that I can't follow you. . ."[31]

But resistance to electrification was not, as this passage suggests, simply a generational difference, but also one of social class.[32] While the TVA rhetorically committed itself to working with the "grass roots," in practice this idea proved vague. Participation "by the people" was assumed to be good, but it was not institutionalized. For example, the agricultural program was not organized at the grass-roots level, but funnelled through land-grant colleges. The result was "a conservative, class-biased" education program where demonstration farms were aimed at prosperous whites. "Black agricultural schools were omitted from the program entirely," and white tenant farmers "received virtually no attention." The TVA allied itself with rather conservative groups, and seldom worked through the Farm Security Administration, the Soil Conservation Service, or the Southern Tenant Farmers Union.[33] The TVA's power resources were never controlled by local municipalities or rural cooperatives. David E. Whisnant concluded that despite decades of "modernizing the mountaineer" by 1965, "it was clear that none of the development efforts ever employed in the Appalachian region had substantially improved the lives of the majority of its people."[34]

V

Like a modernist novel, these stories of rural electrification express contradictory points of view. Multiple narrators have been common in literature since Henry James, but they have found little favor among historians, who prefer to adopt third person, omniscient narration. And I too, while evoking the possibility of multiple narration, still adopt the third person, even in presenting four versions of the same story enfolded within a larger perspective. In the long view, certain clear trends emerge and some points of view turn out to be fallacious. The immediate historical actors—social critics, utility executives, bureaucrats, and farmers—each pursued their own goals, and each has been championed by later spokesmen. This in itself might be understood as an ordinary historiographical problem.

But some recent literary theorists wish to present all texts and events as unstable contestations of meaning, making it impossible to combine these many voices into one historical account. Yet it is not necessary to reduce history of "a rubble of signifiers."[35] No labor of deconstruction can undo the stubborn facts of how many farmers there were, how many had electricity at different times, how much it cost to distribute this electricity, what products farmers purchased, and so forth. If historians cannot pretend to a complete objectivity, nevertheless they can establish a bedrock of facts that permit them to determine limits to what is true and what is false. We can establish such statements as the following to be true: "By the 1920s California had electrified half of its farms as part of its extensive irrigation systems, while less than 5 percent of the farms in Texas had electricity." "Rural customers for electricity were thinly distributed in Montana and the Dakotas." "The percentage of farmers in the population of the United States declined by two thirds between 1950 and 1970." Such matters of chronology, measurement, and census-taking are building blocks of narrative.

We certainly want to know the points of view of all the actors involved, but any valid narrative of American rural electrification is held together by certain over-arching trends. To be acceptable, it must recognize (1) that as electricity came to the land, the farmers were leaving the land, (2) that in the United States electricity remained primarily a commodity, whatever its uses as an instrument of social policy, and (3) that farmers were as consciously a part of the marketplace as the utilities or the government. What emerges in the long view is the primacy of that marketplace. It frustrated the goals of reformers who hoped to see a revival of rural life; it guided businessmen in their development of the electrical grid; it shaped the economic policy of cooperatives; and, on the personal level, it guided the consumption choices made by farmers who decided how to electrify their property. Like it or not, the capitalist marketplace, with all its defects, appears to be the framework for any master narrative we might write about the long-term effects of New Deal rural electrification. Of the 2 percent of the population that remain on farms, roughly two in five are part-time farmers who have other jobs. There are still a good many family farms, but most food is provided by a minority of large farmers with huge investments in land and capital equipment. They rely on satellites for weather information, watch futures markets for guidance on crops, and employ sophisticated electronic measuring devices to monitor their activities. The rural electrification project, even though it focused on small farmers, proved to be a crucial step toward this industrialized agriculture.

The exercise of identifying conflicting narratives has its virtues. Most obviously it facilitates exploration of the points of view of the participants themselves, in part through the recognition that people see their actions as parts of larger narratives. In addition, understanding rural electrification as a bundle of stories helps the historian see New Deal public relations and self-dramatization more clearly. Producing rural electrification as a set of rival discourses momentarily frees us from being a prisoner of any one account. Historians have always needed to be experts at weighing evidence. I am suggesting a modest extension of that practice, the weighing of narratives. It must by now be obvious that my order of presentation has undercut the "plot" of rural electrification as told by the New Dealers and reformers, in order to lay bare underlying market forces that virtually eliminated rural society. Whatever Americans may have thought they were doing, they prepared the way for an electrified neo-rural community which is still in the process of emerging, and which is not primarily based on agriculture.

Figure 4: Electrical Facility *(National Archives, Washington DC)*

FIVE
Energy Narratives

I

"Energy" is a slippery word, referring to something noticeable yet nevertheless intangible. "Energy" may be a noun, but it refers to neither a solid thing like a shoe nor to an abstraction like "liberty" or "democracy." It might be said that this noun aspires to be a verb, since it refers to an active, almost simultaneous, movement from one state to another. Energy is both a commodity and a fundamental aspect of being. Given this peculiar status, energy can function within a narrative explicitly as a prize to be won. But it also belongs to the realm of pre-conditions or choices made implicitly in fashioning a narrative design. This essay deals with both explicit and implicit energy narratives.[1] In the implicit form energy is a central part of the social order which an author assumes to be already in place, and the surface of the narrative may only obliquely refer to these assumptions. In the explicit form a particular technology, such as electric high-tension lines or a hydroelectric dam, becomes the subject of or the location for social struggle, as was the case with William Wister Haines' 1930s working-class novel, *High Tension*.[2] The following survey will look at a suggestive variety of sources, both famous and obscure. Literary works will be placed in a larger context that includes political speeches, advertisements, and public relations, all of which frequently are energy narratives.

Energy has not always been a central category in understanding society or the economy, much less literature. Energy entered narrative once power had spatial elasticity. Before steamboats and railways emerged in the early nineteenth century, energy was only as mobile as a man or horse, and far from being ubiquitous it was scarce and concentrated in only a few places, notably wind and water mills. At the time of the American Revolution there were only three steam engines in the United States. Even in 1838 there were only 2,000 steam engines in the entire country, and the majority of these drove trains or steamboats.[3] In contrast, there were 60,000 small water-powered establishments in 1840.[4] Most energy was local and its origin was immediately identifiable. It did not have the continental dimensions of the railroad, or the instantaneous movement of the telegraph. Most mechanical energy was local and limited, and narratives constructed before c. 1830 necessarily assumed the primacy of human muscular energy and natural forces. In later decades explicit energy narratives became common, as the rapid increase in available power touched directly or indirectly all areas of society and became part of a pervasive ideology of growth and abundance.

The narrativization of energy also was part of a larger shift in scientific thought during the nineteenth century. Until the 1840s scientists worked on the assumption that "force" was the central concept in defining and explaining the universe. But increasingly it became evident to them that "force" was a compound of empirical observation and popular metaphor. Beginning in the 1850s Lord Kelvin and others formulated the laws of thermodynamics, which proposed a new model of the universe in which "energy" was a central term. Only later did their scientific ideas become part of the larger cultural discourse, in a process admirably traced out by Ronald Martin in *American Literature and the Universe of Force*.[5] Energy did not long remain an abstract concept in physics, but was manifested to most people in the particular forms of energy that were becoming central to industrial society, most obviously steam engines and, later, electrical motors and dynamos. Scientists provided successive models of the universe in which energy remained a key term, and by the end of the century many narratives contained an implicit conception of energy, which found expression both in terms of the individual and the society as a whole. The ideologies of human energy included the perhaps inherently tragic Freudian conception of a reservoir with a limited supply that must find an outlet; the capitalist conception that energy can be increased virtually without limits; and Henry Adams's sense that limitless energy would accelerate society toward a certain doom.[6] In short, the concept of energy was intimately connected to the definition of personality and to the sense of history, and it was often present as a tacit knowledge shared between reader and writer.

II

Literary critics have approached narrative in many ways. Some have understood it in terms of a plot structure, isolating particular functions or elements.[7] This was common among the Russian formalists and French structuralists, and using this method one can, for example, delineate the formal elements of the western film or the romance novel.[8] However, as these examples suggest, this approach works when looking at a single genre, but will not be very helpful with explicit energy narratives, because these occur in many different kinds of texts, ranging from advertisements to novels to political speeches. Narratives also can be understood primarily in terms of the techniques used to tell the story, the method of the New Critics. This approach is useful when intensively analyzing a small number of texts, but does not provide a way to classify a large miscellaneous corpus of materials. More useful is Hayden White's meta-critical method based on a synthesis of classical rhetoric, structuralism, and the literary theory of Northop Frye.[9] White's system classifies historical arguments into four modes of poetic discourse, four different explanatory strategies, four plot structures, and four kinds of ideological implication.

mode of emplotment	mode of argument	ideological implication	poetic discourse
Romantic	Formist	Anarchist	Metaphor
Comic	Organicist	Conservative	Synecdoche
Tragic	Mechanistic	Radical	Metonymy
Satirical	Contextualist	Liberal	Irony

This complex classificatory system can be pared down for use in a particular instance, however, because in practice not every conceivable combination of these elements has been used by people promoting or arguing about energy. Most explicit energy narratives emphasize one of the following five preconceptions: (1) natural abundance, (2) artificial scarcity, (3) human ingenuity, (4) man made apocalypse, and (5) existential limits. These five, which I first identified without recourse to White's system, correspond rather well with it and are better understood through it. (1) The narrative of natural abundance is usually emploted as a romance (in Frye's sense of the term). The mode of argument varies but often is either organicist (Emerson) or formist (as with some entrepreneurs). The ideological implication of the narrative for Americans has tended toward a laissez-faire conservatism or anarchism, in which the state is minimized or rejected. (2) Artificial scarcity is usually emplotted as a comedy. It takes a conflict seriously, but the shortage is generally removed in the course of the story. While people with widely varying political views have made use of this narrative, it tends to carry with it the same ideological implications as the narrative of abundance, and indeed the two forms are often found together. (3) The narrative of human ingenuity is a variant of the story of scarcity, but deserves a separate category because the focus is not on a person or an institution that creates scarcity but on a heroic inventor who creates abundance. The focus is less on nature than on technology, with the result that an organicist mode of argument may give way to a mecha-nistic approach in which scientific laws play an important part. (4) Apocalyptic narratives are tragic in their plot, and usually radical in their mode of ideological implication, by which I mean that certain laws and social conditions lead inexorably to disaster. The mode of argument varies usually between either Organicist (nature betrayed by science) or Mechanistic (scientific laws inexorably being worked out), although there are other possibilities. (5) The narrative of existential limits is often a satire, qualifying the hopes for abundance. It presupposes "the ultimate inadequacy of the visions of the world dramatically represented" in the other genres."[10] Any of these five narratives may seem to be true or actu-ally be true at a given time. Often several of them simultaneously compete to explain events and to suggest various ways to deal with them. My purpose here (in contrast to the preceeding chapter) is not to determine their veracity, but rather to identify the major forms and provide examples of how they have been used.

The narrative of abundance emerged with the industrial revolution, as the muscle power of human beings, draft animals, and simple water mills was exceeded by new machines. In this narrative more energy continually becomes available, assuring profits, progress, and personal success. This became the central energy narrative for the United States, against which all the others were defined. The assumption that energy is naturally abun-dant is especially suited to a laissez-faire ideology, in which the self-reliant individual has only to make use of his own energies in order to rise in the world. This story was vigorously promoted by Alexander Hamilton and early industrialists, and it flourished in the nineteenth century. In popular culture it often focused on the railroad, both as an object of power and as a force expressing American culture. It was embodied in a famous Currier and Ives 1868 lithograph, "Across the Continent," in which a railroad train pushes out across the great plains.[11] Ahead the track is still being laid, but the area already served by the railroad contains a prosperous new community, with a prominent public school. The power to move across the land is translated into expansion and settlement, while Native Americans and wild animals retreat into the distance.

During the twentieth century corporate public relations has been the most obvious disseminator of this narrative. Roland Marchand concluded that advertising in the 1920s and 1930s used "scores of parables and visual clichés" to beckon "consumers to join in a cost-free progress toward modernity."[12] General Electric, for example, designed a campaign to create an "electric consciousness." In addition to spending millions of dollars on advertisements, the company also disseminated this message through personal appearances by Charles M. Ripley, a popular lecturer. The indefatigable Ripley traveled the United States repeating the same talk, "The Romance of Power," often giving it more than once a day. Like a barn-storming revivalist he spread the gospel of progress through electricity, speaking to Rotary clubs, schools, women's groups, and any other organization where local utilities could introduce him. So popular was his lecture that the 100 slides were reproduced in sets and other men were trained to give the same talk. In 1928 the presentation was repackaged as a book, *The Romance of Power*, and distributed to schools all over the United States. It had the following subtitles: "Travel, Life and Labor, Here and Abroad;" "An Illustrated Review Filled with Laughs and Information;" "The ABCs of Human Progress;" and "Pictures Showing What the World Can Learn from the United States and What We can Learn from the World." As might be expected, the rest of the world had but little to teach, but could learn a great deal from America. Photographs showed "primitive" people doing things by hand that Americans did with machines. Ripley had statistics and charts to prove that Americans had a higher standard of living than the English, and he argued that the measure of success was the use of energy. British factories used only one-third as much horsepower per worker as American factories, and, he claimed, productivity was only one-third as great. Ripley concluded his book with the admonition, "Young men and women, guard this mechanism [private power plants]. Keep the wheels turning, faster and faster, making more and ever more of the good things of life—for everybody."[13]

Such sentiments were widely held. They could be found not only in corporate advertising and public relations, but just as importantly they were embraced by a whole spectrum of political figures from conservative Americans to Lenin. As explored in chapter seven, corporate exhibits at the world's fairs in Chicago (1933–1934) and New York (1939–1940) proclaimed how private ownership of energy would lead to abundance.

Likewise, as discussed in the previous chapter, the politicians who funded the Tennessee Valley Authority (TVA) were certain that they were creating abundance. Even at the political extremes electricity was seen as a catalyst for progress. Both the conservative Southern Agrarians and Marxists of the 1930s believed that increasing electrification was virtually identical with prosperity and progress. The narrative of abundance was dominant from at least 1830 until 1960, and has by no means lost its popularity. It is a totalizing vision, to which all of the other energy narratives refer. During the 1950s an updated version of Ripley's talk was purveyed by another General Electric public relations man, Ronald Reagan.

The second form of explicit energy narrative, like the first, assumes that an abundance of energy exists, but it focuses on what are perceived as artificial limitations of the supply. This is usually a melodramatic tale of good and evil, in which natural abundance is cut off or hoarded for personal gain. Before 1940 this story was commonly found in novels that attacked private monopolies of electrical power.[14] But there have been many variants of this narrative. Some have depicted public utilities as an inherent evil, claiming that compared to private enterprises they were so badly run and inefficient that they hampered development.[15] Seen in this perspective, Theodore Dreiser's Cowperwood novels concern artificial limitations on energy and the hero's attempts to rationalize its use. This narrative could also be used to attack miners if they cut off the supply of coal, as was the case during the great anthracite strike of 1902 in Pennsylvania.[16] More recently, during the energy crisis of the 1970s, the American Electric Power System ran full-page advertisements advocating greater use of American coal, "our ace against Middle East Oil."[17] In their view, misguided and over-zealous regulation by the Environmental Protection Agency prevented mining of eastern, bituminous coal while the Department of the Interior refused to allow access to "millions of tons of clean western coal."[18]

In yet another variant of the narrative of limitation, short-sighted technicians could inadvertently create blockages in the free flow of energy, if no one was ready to consume it. For example, a 1940s novel dealing with the TVA focused on the difficulties encountered in evacuating backwoods people from their ancestral homes and educating them to accept modern ways, in order to make way for progress. As one character lectures an engineer, whose only concerns are technical, "If you were allowed to build this dam unhampered by the rest of us, it might provide electric power, but for whom? A malaria-ridden, poverty-stricken, one-crop population farming burnt-out land can't buy electricity, nor can it buy the products of factories. . . . The more highly you impractical technicians industrialize a region, the wider becomes the gap between the articles you produce and the consumers on whom the ultimate success or failure of your enterprise depends."[19] More commonly engineers were depicted as the heroes of the abundance story. Well before World War I, the engineer emerged as a rugged individualist, the man of action (the profession was overwhelmingly male) who overcame obstacles, put theories into practice, and ensured material progress.[20]

While early limitation stories focused on the US, since World War II the narrative of artificial or unnecessary limits has assumed global dimensions. Frequently it is associated with foreign policy, national self-interest, and the deployment of troops in the Middle East. In the 1970s OPEC version, the narrative focused on conflicts between the United States and the cartel. In 1990, when Saddam Hussein invaded Kuwait, he was easily type-cast in the role of the villain who threatens the world's energy supply, first by seizing oil wells and later by setting hundreds of them on fire.

The third form of the explicit energy narrative is a tale of transformation, in which clever technicians reveal how to achieve growth, progress, and personal success by discovering new resources or recycling old ones. This story is especially welcome during times of shortages. In 1974, during the oil boycott, almost 5,000 individual publications in all areas of the American trade press carried special advertising about how to conserve energy, much of it prepared on a volunteer basis by major advertisers.[21] But this narrative is hardly new. Henry Ford had expressed most of its basic ideas by the early 1930s. He early championed recycling, the use of renewable resources and non-polluting forms of power generation.[22] Ford also helped to enshrine Thomas Edison as one of the central figures of this myth, in which the scientist or inventor is the hero.[23]

After 1940 atomic power began to be presented in terms of the transformation narrative. As a fuel thought to be cheap and inexhaustible, it underlay visions of a future world whose energy supplies would be so limitless that crops could be grown with artificial light underground, leaving the surface of the earth free for recreation. One scientist wrote in *Collier's Magazine* in 1940 that atomic energy would ensure "unparalleled richness and opportunities for all. Privilege and class distinctions and the other sources of social uneasiness and bitterness will become relics because things that make up the good life will be so abundant and inexpensive." The same year, *The Saturday Evening Post* called uranium "a miraculous new continent of matter, as rich and wonderful in its way as the Americas proved to be years after their discovery," and spoke of "the Promised Land of Atomic Energy."[24] A generation later, economist Robert L. Heilbroner stressed that, "Given enough power, which nuclear energy now begins to promise us, we could literally 'melt' the rocks and reconstitute any substance by synthetic processes. . . the long-term future holds out much more promise than the anti-growth school of thought reveals."[25] (In the same vein, the apparent discovery of "cold fusion" generated a considerable excitement at the end of the 1980s.) While such heady visions have receded, this narrative remains common. More recent versions stress the development of new energy sources such as solar power to replace fossil fuels when they either run out or become too expensive, the adoption of more energy-efficient housing designs, and the installation of low-energy appliances.[26] In this story there are no immediate limits to growth if Americans use appropriate technologies.

In some versions of the transformation narrative, mastering energy through technology leads to human transformation. Edison speculated that the introduction of lighting into communities not only kept people awake longer but quickened their intellectual and social development.[27] Matthew Luckiesh, one of General Electric's leading lighting researchers during the 1920s and 1930s, tried to document similar theories scientifically.[28] In the 1980s Murray Melbin, a sociologist, expressed a related view in *Night as Frontier: Colonizing the World After Dark*. Melbin argued that people in post-industrial societies live in a fundamentally new way, because they no longer obey the diurnal rhythms of night and day, but continue around the clock without interruption, creating a new social environment characterized by "incessance." He concluded:

> Our characters and our bodies are changing and becoming suited to the particulars of incessance. Being reared in a more wakeful household, accelerating the timetable of maturation, hormonal levels in the blood being less conducive to drowsiness, and practices of repeatedly recombining in personal relationships, all contribute to a better fit between persons and nighttime undertakings. Social, biological, and psychological processes have combined to transform us in ways to fit the environment we refashioned, and the community is being stocked with people who are more comfortable with its timetable. Along with altering our milieu, we are altering ourselves.[29]

In contrast to this evolutionary optimism, the apocalyptic narrative emphasizes the destructive force of energy sources. While the origins of this story can be traced back to British fears of "satanic mills" in the late eighteenth century, similar dystopian expressions became common in the

United States after the Civil War. Mark Twain's *Connecticut Yankee in King Arthur's Court* apparently begins as a narrative of abundance. A nineteenth-century mechanic who has been transported back in time vows to bring industrial civilization and all its benefits to early medieval England. But the inventions that have the greatest social impact turn out to be not the bicycle and telephone, but gunpowder, electric fences, and the machine gun. His adventure ends with a battle between high technology and traditional weaponry, in which thousands of knights are electrocuted or slaughtered.[30] Twain's contemporary Henry Adams extrapolated energy development into the future, and concluded that historical events were accelerating out of control.[31]

Such fears were given concrete form by the atomic bomb. After Hiroshima many apocalyptic novels and films questioned the consequences of energy development. The 1954 Japanese film *Godzilla* imagined a reptile that had become radioactive from bomb tests in the Pacific and then had grown so enormous it could crush the buildings of Tokyo beneath its feet. In the same year an American film *Them!* terrified moviegoers with another mutation: ants the size of trucks, spawned by tests in the desert. A similar film showed an army of giant grasshoppers attacking Chicago.[32] A more realistic 1950s apocalypse was *On the Beach*, which depicted the gradual eradication of mankind by nuclear fallout. Set in Australia, the film shows how the last human beings on earth cope with their fate, as deadly winds bring the fallout nearer. Many commit suicide. Other visions of nuclear destruction appeared regularly in succeeding decades, including *Dr. Strangelove*, *Fail Safe*, and *The China Syndrome*. The public was constantly reminded that both scientists and the systems they designed might get out of control. After the early 1960s the "most typical image" of the world after a nuclear war was of "one boundless Hiroshima."[33]

By comparison, the fifth form, the existential narrative of absolute limitation, is bleak but ends with survival. It does not blame human beings for mistakes or deficiencies, but presents energy scarcity as a fact of nature, based on scientific principles. The first law of thermodynamics is that the total energy in the universe always remains constant, but the second law (that on entropy) states that whenever any conversion of energy takes place, some energy becomes unavailable for future use. These laws led Lord Kelvin, who helped formulate them, to predict the gradual heat death of the universe. Such speculations encouraged a strain of romantic pessimism in popular writing, particularly in science fiction. Often a central figure must come to terms with limits, such as the computer whiz kid Fisher in Paul Theroux's *O-Zone*.[34]

In some novels capitalists are forced to recognize the limits of expansion. As early as the 1930s, *Bloodbird*, a forgotten work by Thomas Burton, pits the Winthrop family, which plans to build a hydroelectric dam, against "the valley farmers" whose lands will be sacrificed to the project.[35] Melodramatic narratives of artificial limits had been concerned primarily with reforming the city; Burton suggests that the still unredeemed city could threaten the well-being of the countryside as well. The utility buys out some farmers and forces out others through bankruptcy proceedings. One by one they are beaten, until only two remain, while dam construction devastates the valley. *Bloodbird* inverts the abundance narrative, emphasizing the human and natural costs of power generation. In this world of natural limits the farmers are associated with the idealized Jeffersonian world of subsistence farming and individualism. They look back to the pre-industrial period for values and standards, and ultimately the novel will vindicate their point of view.[36]

As this example suggests, the narrative of limits has long been an alternative to the Hamiltonian vision of unlimited expansion, and can be traced back to Thomas Jefferson, who argued that the United States should remain an agricultural nation and minimize both urbanization and industrialization.[37] Similar narratives are common in much of the literature of ecology that emerged in the 1970s. They often focus on the hazards of poor design and the limits to technological growth. Ernest Callenbach dramatized this view in *Ecotopia*, a novel which imagines an independent society located in the Pacific Northwest.[38] After breaking free of the rest of the United States, it has introduced high-speed electric rail everywhere, abolished automobiles, and turned the centers of cities into quiet residential areas. Consciously halting growth and accepting a lower gross national product, it has achieved an ecological balance with the environment. Many areas have been reforested, and the educational system has been revamped to emphasize ecological awareness. Low-energy houses and appliances have been introduced, not with the goal of permitting more growth, but in order to achieve a steady state. Callenbach envisioned a world where E. F. Schumacher's *Small is Beautiful* had become political orthodoxy. Published in 1973, two years before Ecotopia, it also proposed downsizing consumer demands and the scale of technologies.[39]

The explicit energy narratives that emerged in the twentieth century both responded to current events and projected resolutions to cultural contradictions. During the two energy crises of the 1970s all five of the narratives discussed here could be found in the public discourse. Coal companies emphasized natural abundance. Oil companies complained of

artificial restrictions, often combined with claims for how their technological ingenuity could overcome the crisis. Environmentalists fashioned either
apocalyptic or existential narratives. The public as a whole lacked the expertise to evaluate these stories as recommendations for public policy or as
scientific claims, but they were able to judge whether a scenario seemed
convincing. By 1980, a majority was persuaded by Ronald Reagan's
rearticulation of the oldest and most familiar narrative: natural abundance.

III

Perhaps because Jimmy Carter was the only recent president who knew
much about the laws of thermodynamics, he often described America's
energy future in the terms of the narrative of inescapable limits. Less
frequently he presented voters with stories of technical transformation,
emphasizing the long-range development of alternate forms of energy. But
with his training both as an engineer and as a nuclear submarine
commander, Carter saw clearly the problems with nuclear power, and he
also realized that alternative energy sources such as wind and solar power
would not soon be available on a commercial scale.

During the second oil crisis Carter turned to the existential narrative.
In 1977 he declared, "The energy crisis has not yet overwhelmed us, but
it will if we do not act quickly. It is a problem we will not be able to solve
in the next few years, and it is likely to get progressively worse through the
rest of this century. Our decision about energy will test the character of the
American people. . . [it] will be the moral equivalent of war."[40] Sitting in
the White House with the heat turned down, Carter exhorted Americans
to engage in a decades-long fight to overcome their own weakness (over-
consumption) and dependence (on foreign oil). The possible truth of this
narrative appeared to be confirmed by the long lines at the gas pumps, the
rising cost of oil, and the simultaneous stagnation and inflation in the
American economy.

But this narrative of existential limits has never been widely popular.
And in an election year the idea of an overwhelming, long-term crisis and
sacrifices, dictated by limits in the energy supply did not play well with the
electorate. Candidate Ronald Reagan campaigned successfully against
Carter's scenario of suffering. In his acceptance speech at the 1980
Republican National Convention he refused to acknowledge that energy
shortages even existed. Instead, he redefined Carter's program as a narrative of artificial limitation, and declared that over-zealous regulation had
strangled the free market.

Those who preside over the worst energy shortage in our history tell us to use less, so that we will run out of oil, gasoline, and natural gas a little more slowly. . . .But conservation is not the sole answer to our energy needs. America must get to work producing more energy. The Republican program for solving economic problems is based on growth and productivity. Large amounts of oil, coal, and natural gas lie beneath our land and off our shores, untouched because the present administration seems to believe the American people would rather see more regulation, more taxes, and more controls, than more energy.[41]

During the rest of the campaign Reagan continually repeated: "America has an abundance of energy. But the policies of this administration consistently discouraged its discovery and production." He told two stories: the narrative of abundance, emphatically associated with America's past and future promise, and the narrative of artificial and unnecessary present shortages imposed by Democrats. He promised to "get America producing again," and conflated Carter's policies with those of the oil cartel, without explicitly saying so, by depicting both as parts of a narrative of limitation.[42]

Reagan's preferred narratives (abundance and artificial limits) contain an almost transcendental vision of technology as a benign human intervention in the natural world. In contrast, the explicitly technological narratives (those of human ingenuity, apocalypse, and existential limits) recognize a tension between nature and industrialization. These narratives seek to resolve that tension either by increasing the energy that can be extracted from the same resources, or by reducing demand. While such narratives began to appear by the end of the nineteenth century, they were not taken seriously by a large public until the 1970s. All three technological narratives were well represented in debates spurred by the Club of Rome's publication of *The Limits to Growth*, and they still remain prominent in debates on global warming and deterioration of the ozone layer.[43]

IV

We have seen how contrasting energy narratives provide scenarios for social and economic decisions. Yet this typology hardly exhausts the subject of energy narratives. Indeed, it has a mechanical feel about it and is too static and *a priori*, treating energy as an economic and political factor and understanding it as an object, rather than seeing it as an underlying dynamic force that takes various forms in different historical periods. To understand energy as a factor in literary works it must be grasped as a changing category of daily experience, not only as an object of political debate. In most literature energy is not a possession to be fought over, nor do authors conceptualize it explicitly. To discuss energy in relation to literary form, the scheme presented so far is inadequate, because it suggests that narratives neatly correspond with fully articulated ideological positions. In practice, energy often was implicit, especially as electricity supplanted steam power.

Naturalism as a literary mode seemed particularly appropriate to the steam engine, while the more complex literary techniques of modernism emerged along with the full electrification of society. What is the connection? During the steam era, in these two modes of production, literary and industrial, all the moving parts seemed clearly visible. A naturalist narrative, with its characters linked together in a sequence of actions tightly controlled by heredity and environment, resembles the steam-driven factory, with its clearly articulated system of belts and gears. The world of the steam-driven factory may have been large, but it was technologically limited in size. The shafts, belts, and gears could only extend so far. The linkages were visible and the energy system was transparent to all observers. It was a system of unambiguous cause and effect, where the relations between owners and employees were clear.

In contrast, the electrified factory expressed a modernist aesthetic, not only in its elongated, functionalist architecture, but in its interior articulations of power. Instead of a central power source from which emanated all the drive shafts, the new energy source was flexible and invisible, and it could be transmitted anywhere. The form of the factory was no longer limited by the capacity of shafts and belts to carry power over distances. Rather, the form became infinitely flexible, as all spaces became equally accessible to power, and the factory could be expanded over a much greater area. This (re)construction had profound consequences for the imagination of social relations. What had previously been remote or inaccessible now potentially came under control. As power expanded so did the intrusiveness of those who controlled it, as Charlie Chaplin brilliantly suggested in the factory scenes of *Modern Times*.

Yet at the same time that the new power system was ubiquitous it was elusive. The changes in space and scale were facilitated by the interposition of telephones, loudspeakers, and other electrical communication devices between managers and workers. Most important of all, workers ceased to experience manufacturing as a series of coherent steps leading to a finished product that they understood from start to finish. Instead, they repeated a single task endlessly, as modern work required simultaneous work on everything at once. The factory's homogenization of space and simultaneity in time were likewise the conceptual underpinnings of much modernist writing. The shift from one energy system to another, from steam to electricity, provided a whole new conceptual and experiential framework.[44]

Implicit energy narratives during the 1920s registered this shift from one energy system to another. Pierre Macherey wrote about such often unacknowledged transformations in *A Theory of Literary Production*:

> the [literary] work is articulated in relation to the reality from the ground of which it emerges: not a 'natural' empirical reality, but that intricate reality in which men—both writers and readers—live, that reality which is their ideology. The work is made on the ground of this ideology, that tacit and original language: not to speak, reveal, translate or make explicit this language, but to make possible that absence of words without which there would be nothing to say. We should question the work as to what it does not and cannot say, in those silences for which it has been made. . . . The order which it professes is merely an imagined order, projected on to disorder, the fictive resolution of ideological conflicts. . . . The work derives its form from this incompleteness which enables us to identify the active presence of a conflict at its borders. [45]

What then are the ideological conflicts created by the shift from one energy system to another? And how do these conflicts create the uneven ground of literary experimentation?

Consider the deployment of millions of lights in Times Square, and the transformation of the city into an electric landscape. The Great White Way was a universe of signs that proclaimed not only particular mademade products but the creation of a new landscape. For the millions of tourists who came to stare at them in Times Square, the signs only incidentally advertised an array of products. They came to see the sheer size and magnificence of the flashing signs; they were engulfed in a restless crowd and the roar of the city. This electric landscape, even more than the new electrified factories, was the cultural ground from which modernism sprang. From one point of view its presence signified the standardization of products, the use of advertising as a means of mass persuasion, the power of large corporations, and a widespread celebration of technology. But for the artist and the writer, the landscape's chaotic brilliance also expressed an implicit ideology that valued simultaneity, fragmentation, and montage. This new electric landscape stamped itself upon the imagination, and became a central part of the intricate topography of modernist experience. It remains, then, to examine one well-known text to see how this analysis might prove useful.

V

The Great Gatsby is an exemplary implicit energy narrative, for Gatsby uses electricity as a tool of self-creation.[46] Indeed, one of Carraway's first attempts to describe him is as an electrical machine capable of sensing seismic disturbances. Gatsby controls his bootlegging and illegal bond sales using electrical technologies, the telegraph and telephone, which permit him to keep track of far-flung operations while appearing to be a rich man of leisure. These technologies allow him, in effect, to be simultaneously in several places at once, which is to say that he can be several persons at once. The butler need only call him away to the phone for a moment, and he enters one of the silences of the text, the unheard conversations with his underworld connections. It is hardly a flaw in the novel that we only hear a fragment or two from these phone calls. What we do hear establishes what sort of activities Gatsby is monitoring and directing. These silences only hide the specific content of his illegal activities; they are the surface beneath which he can devise multiple identities.

Nick encounters these many identities at the lavish parties held on Gatsby's spacious lawn, where his servants erect "enough colored lights to make a Christmas tree of Gatsby's enormous garden." The lights transform the home's appearance, inviting guests to attend the parties as though they were at Coney Island or on Broadway. The guests include actors, producers, and singers from the entertainment world. The elaborate party lights are not only Gatsby's advertisement of himself to the world at large but they create a world where illusion and reality blend. At the same time, these lights, like those on the Great White Way, are an advertisement sent across the bay to Daisy. The lights proclaim him to her, and in his imagination the green light at the end of her dock is a winking response.

On one occasion Gatsby puts on all the lights in his house to "glance into a few rooms," as if by making them more intensely visible his success will become more real. The closer Gatsby comes to regaining Daisy, the more intensely they shine, culminating on the night when he has finally arranged a meeting, when they are at their brightest. Carraway recalls, "Two o'clock and the whole corner of the peninsula was ablaze with light, which fell unreal on the shrubbery and made thin elongating glints upon the roadside wires. Turning a corner, I saw that it was Gatsby's house, lit from tower to cellar."

This was to be his last great display. After he has recovered Daisy, Gatsby ceases to exaggerate himself through electricity, and attempts to retire into a private life. With Daisy by his side he no longer needs to hold the parties, and "the lights in his house failed to go on one Saturday night—and, as obscurely as it had begun, his career as Trimalchio was over." In effect, Gatsby wishes to give up his split existence, divided between social pretense, half-confidences to Carraway, deals with the underworld, and a secret romantic love for Daisy. At the book's outset he had achieved what might be called a cubist personality, which Carraway can see from many perspectives. Once he regains Daisy, however, he seems to want a single whole identity, like a character in a realist novel. He hires new, unfriendly servants that ward off the public, and he tries to retire from a public role. Of course Gatsby cannot, as Carraway tries to tell him, turn the clock back to an earlier time. He cannot become an unambiguous character in the world of realism, with its gears and belts of cause and effect that ensure a logical plot. Instead, his identities become entangled with Tom's deceptions (which also rely considerably on the electrical technology of the telephone). Wilson kills Gatsby in what might be called a case of mistaken identity, but which more accurately is an outcome of multiplying identities and self-deceptions. Wilson is doggedly naturalistic, a man with only one identity. Gatsby deploys himself as a multiple character, based on the simultaneity, pretense, and social display that electrical technologies make possible, generating endless rumors about himself. Because he is fragmented, he can make intermittent and startling appearances in the novel, like a great advertisement blinking on the Great White Way. He is entertaining for a season, but soon is emptied of content, burns out, and must be replaced.[47]

As *The Great Gatsby* demonstrates, novels which do not appear to have anything to do with energy may register the shift from one system to another. They can contain the tension between two systems of values, experiences, and metaphors, the one based on steam power, cause and effect, the gears of the clock, and self-control, the other based on electrification, simultaneity, and a self in fragments. In early life Gatsby attempts to shape himself into a Franklinesque character who can make his way by force of character, but this attempt is doomed to failure in a discontinuous, electrified world. Likewise, the novel's famous "wasteland" landscape between Long Island and New York takes on additional resonance when understood as part of an energy narrative. Literally, the ashes are a pointed reminder of the wastefulness of the high-energy society, the entropic result of burning coal. Looking at this landscape, Wilson sells energy at his gas station; he is always tired, as lacking in energy as the grey world of ashes. Figuratively, this borderland isolates him and his nineteenth-century notions of personal advancement, and it mocks his demand for a literal cause-and-effect explanation, or a motive, to explain the death of his wife.

Nick Carraway presents quite a different case. He has given up on the old universe of force and self-creation, which for him is located in the past which he evokes on the last page of the novel. Only when the electric lights have been turned off along the shore can he imagine the inexpressible wonder of Dutch sailors who looked out at the fresh green breast of America. Projecting himself into this abstract space which has not yet become a landscape, he begins to write the old story, imagining the promise of abundance in an orgiastic future, symbolized by a green light. But realizing the impossibility of this older narrative, his text breaks off in an ellipsis, falling back into the silence from which it has been made, a silence imagined as the unelectrified space of the earlier republic, the lost space of possibility.

SIX

Space of the Past: E. L. Doctorow's
World's Fair

I

The most common ways of knowing the past are scholarly histories, literary works, and reconstructed sites. The boundaries between these "ways of knowing" have never been firm, and in recent decades they have become increasingly blurred. The "non-fiction novel," documentary film, oral history, historical restoration, and many other forms defy traditional classifications. At the same time the narrative strategies of postmodernism have challenged the form of history writing itself, which has generally clung to its own version of nineteenth-century realism. The common element in all these forms of representation is narrative. In *Metahistory* Hayden White found three principal narrative strategies shared by historians and novelists but which also could be extended to popular culture: explanation by emplotment, by argument, and by ideological implication.[1] White's theory proved effective in an analysis of major nineteenth-century historians. His larger project, carried forward in a series of influential articles and books, has had the effect of blurring the line between fiction and history, raising the question of whether there is any fundamental difference between them. E. L. Doctorow's *World's Fair* provides the occasion for a meditation on this question, since it exists on the borderline between autobiography, history, and fiction, and it incorporates a great deal of the popular culture of the 1930s.[2]

White refuses to make qualitative judgements between his historians, treating them all as equals, but these "equals," who do not agree in their interpretations of the past, are all eminent, such as Ranke and Tocqueville. They have achieved a reputation for dealing accurately and fairly with documents. In comparing them White has argued that the difference between interpretations of events, and hence the meaning of one history as compared to another, arises through the techniques of encoding facts as parts of a larger design.

> . . . the primary meaning of a narrative would then consist of the destructuration of a set of events (real or imagined) originally encoded in one tropological mode and the progressive restructuration of the set in another tropological mode. As thus envisaged, narrative would be a process of decodation and recodation in which an original perception is clarified by being cast in a figurative mode different from that in which it has become encoded by convention, authority, or custom. And the explanatory force of the narrative would then depend on the contrast between the original encodation and the later one.[3]

White argues that given the same information historians nevertheless arrive at different interpretations, even though they may agree on such basic matters as chronology, which documents are legitimate and which false, and so forth. He focuses upon how differences arise though variation in the strategies of representation.[4]

Doctorow's often quoted claim—"There is no fiction or nonfiction as we commonly understand the distinction: there is only narrative."[5]—is apparently in agreement with White's position. And when such statements are coupled with Doctorow's repeated literary raids on history and the popularity of these works, one might ask if the wider culture has not arrived at the same conclusion as White. Doctorow would almost seem to be a new kind of historian. Certainly, he has done background research for his novels. To take one small example from *Ragtime*, Henry Ford did believe in reincarnation, though this fact is hardly well known.[6] However, in the novel this fact is only briefly introduced as part of a fabricated meeting between the automaker and J. P. Morgan. Ford's religious views may be amusing when treated in this way, but they are not meaningful until made part of some larger pattern. Ford's interest in reincarnation, Transcendentalism, thought-transfer, and the New Thought movement is only a curiosity until linked to his Model-T car and the assembly line. The novelist's contribution to history can be to take such facts and explore the possible connections between them, going beyond but not violating the evidence that does exist. As Cushing Strout put it, "Classically, the historical novel finds its opportunity in exploring matters that historians have left unsettled for lack of conclusive evidence."[7]

Doctorow engaged in such a procedure in *The Book of Daniel*. He began with a set of contradictory facts concerning the Rosenberg case, and modified them in some ways which were not crucially important and developed a new way to understand it. As Strout argues, such a novel exemplifies the veracious imagination at work; in contrast the imagination at work in *Ragtime* had became voracious, devouring the past. Strout's comment: "To make historical characters do whatever suits one's fancy makes for amusing entertainment or for political manipulation, but the veracious imagination is corrupted."[8] In making this distinction, Strout seems to imply that veracity, meaning factual accuracy and avoidance of anachronism, is all that history requires. However, while *World's Fair* can fairly be called a product of the veracious imagination, it is too incomplete to be history, and its selection and presentation of historical material are questionable.

The book's weaknesses are linked to its form, which limits the imaginative work it asks the reader to perform. As an example of the kind of imagination that reading Doctorow calls for, consider a brief passage from *Ragtime*. "He computed the cost of all the fares. It would come to two dollars and forty cents for him, just over a dollar for the child. The trolley hummed along the dirt roads, the sun behind it now going down in the Berkshires. Stands of fir trees threw long shadows. They passed a single oarsman in a scull on a very quiet broad stream. They saw a great dripping mill wheel turning slowly over a creek."[9] The passage is factually correct; it was possible to take an electric trolley trip from New York to Boston, as Tateh and his daughter do. Doctorow even has roughly correct information on the fares. In *World's Fair* he also writes such spare short sentences, balanced between the interior experience of a character and a seemingly objective description of the exterior world. The words seem well grounded. To anyone familiar with western Massachusetts the passage evokes particular places, and it can be located along the line of towns stretching from Stockbridge and Lee eastward to Springfield. The "very quiet broad stream" could well be the Connecticut River. The mill wheel is a quaint survival from the early period of American industrialization. As someone who once lived in that area, I can fill in blanks in the prose. Similarly, the prose of *World's Fair* evokes a great range of meanings for the large audience that knows the popular culture of the 1930s. The book triggers whole sets of memories, particularly for people from the New York area, but also for anyone who recalls "The Shadow," "The Green Hornet," Flash Gordon, or Benny Goodman. The clean surface of the writing hides many inviting gaps, encouraging the reader to project personal knowledge and experience, so that the book resonates with associations.

All fiction invites the reader to imagine, of course, but the objectives vary. For a suggestive contrast, recall how James Agee, in *Let Us Now Praise Famous Men*, makes exhaustive lists of the contents of sharecroppers' homes, seeking to make us feel the weight and texture of their lives through a baroque, all-encompassing language. Agee was trying to evoke the experiences of those who did not share a common popular culture with the middle-class reader. For him there could be no easy references to "The Shadow," the New York Subway, or Far Rockaway Beach. Agee sought to overcome the radical differences between the reader and the sharecroppers, and to do so without debasing them or reducing them to historical types. He recognized his separation from the people he wished to describe. In *World's Fair* Doctorow operates from an opposite set of assumptions: the narrator is presented as being typical of his time, and he is given few individualizing features. He uses the common speech, with virtually no distinguishing ethnic or regional expressions. The places he inhabits seem much like all other houses, and the content of his imagination comes largely from popular culture. The book thus becomes a kind of representative life.

Such a work is far removed from the project of White and other poststructuralist thinkers, who for quite different reasons have eroded the distinction between history and literature. *World's Fair* is part of a parallel but not identical process at work in American popular culture, in which the past is increasingly understood through representations in film, on television, or in reconstructions of the "original" Plymouth Plantation, Colonial Williamsburg, or "western ghost town." In many cases the individual elements of these recreations are accurate, but the ensemble as a whole transforms the past through selection, deletion, and emphasis, in much the same way that a writer can select and organize his materials. Every individual fact may be true, but the ensemble as a whole is profoundly false because of what is left out. The popular acceptance of these celluloid visions and idealized towns as an accurate rendering of the past sharply contrasts with White's project, which explores the consequences of being ineluctably cut off from re-experiencing the past by the nature of language and representation itself. While White posits our radical alienation from the past, Doctorow allies himself with the apparent realism of historical recreations and film that claim to make the past accessible. What form and what voice can be the vehicles for this intention?

II

Unlike *The Book of Daniel* or *Loon Lake* which were more obviously experimental in form, at first glance *World's Fair* seems to be a memoir. Indeed, newspaper reviewers often described it not as fiction but as autobiography, suggesting that in the public mind it has been accepted as a personal history. *The Los Angeles Herald–Examiner* described it as, "A wonderful sort of time capsule that recreates the experiences of his childhood and evokes the emotions he experienced." *Newsweek* praised, "Exquisitely rendered details of a lost way of life in New York fifty years ago." Many reviewers construed the book as a blend of autobiography and history, and the publisher encouraged this idea by placing all of the above quotations at the front of the book.

Furthermore, Doctorow suggested that the work is a recollection of his childhood, by placing on the first page an inscription from "The Prelude," which Wordsworth described as a "poem on the growth of my own mind."[10] Doctorow noted in an interview that "the basic convention of *World's Fair* is that it is memoir. That is what it pretends to be in the voice of the protagonist."[11] He continued, "In *World's Fair* I gave the young hero my name, Edgar, but I don't think he's me. . ." By his own account, then, he was not writing autobiography, but rather a representative boy's life and a pastiche of the time period. He sought to blur the line between public and private, fusing the experiences of one child and the popular culture of the 1930s.[12]

Nevertheless, Doctorow's biography is quite similar to that of his principal narrator. Both are named Edgar, both have an older brother named Donald, both grew up in the New York area during the Depression, and both have a mother named Rose. This child narrator has a curious voice that speaks a child's mind and yet paradoxically commands the rich vocabulary of maturity. One of the subjects here is the emergence of the writer, signaled in this case when the fictional Edgar wins an essay contest. Indeed, either the fictional Edgar or the real Edgar might have said, "I really started to think of myself as a writer when I was about nine."[13] Yet an equally important subject is the "typical" boy emerging into a separate identity at the end of childhood. He embraces the artifacts of popular culture, yet he lovingly buries them in a home-made time capsule. Symbolically, just before sealing the capsule he removes from it a popular book on ventriloquism, and the work closes as he walks, "headed into the wind" while he practices "the ventriloquial drone."[14] This ending is particularly suggestive, because Doctorow has changed his voice in each of the novels he has written. In an interview he once defined the problem of writing as being one of determining the right voice to use. Thus the pamphlet on ventriloquism that Edgar acquires in order to master the techniques of appearing to speak through the mouths of others is an obvious metaphor for Doctorow's authorship. Edgar becomes a ventriloquist, a fabulist, a shape-shifter.

In earlier works, such as *The Book of Daniel* and *Ragtime*, Doctorow subverted the usually well-demarcated line between first and third person, between interior meditation and the description of the world. The effect of that narrative technique in *Ragtime* was to elide the personal and the general, making one family's experience into a microcosm of the society. *World's Fair* has a similar aim but uses different stylistic devices, relying primarily on one narrator, plus a few oral histories from family members. Edgar's story is central; the others appear only briefly, giving historical and family background. Doctorow placed these brief interpolations in the text to mimic oral history as a "way to break down the distinction between fiction and actuality."[15]

What is the value of a document that attempts to break down the distinction between "fiction and actuality"? Its depiction of the 1930s contains little original or new, as Robert Towers concluded in *The New York Review of Books*: "The material is familiar from a dozen novels, from books on the Depression era, and from memoirs of growing up Jewish in New York."[16] The book cannot be considered a ground-breaking historical work. Rather *World's Fair* fills a void created when academics ceased to write a certain kind of light, popular narrative. Engaging history, once the norm, has all but disappeared, replaced by a monographic tradition of rigorous argument, massive documentation, and narrow focus. Little wonder that the general reader who could once enjoy Frederick Lewis Allen's *Only Yesterday* now turns to *World's Fair* to get a pleasant, if edited, account of the past. For, as Towers also noted, Doctorow includes "so much observed period detail that a reader who has lived through the thirties will experience repeated tremors, if not shocks, of recognition. The trivia of those years is lavishly spread on nearly every page."

Writing unfootnoted popular histories is a venerable tradition, stretching back at least as far as the *historiens galants* of the seventeenth century, who composed stories of intrigue in a belletristic spirit.[17] In *World's Fair*, unlike *Ragtime* where anachronism and improbabilities abound, Doctorow has observed its conventions. He carefully observes chronology and his facts are subject to verification: the reader can check the names of radio shows and their sponsors, or find out if phonographs were built as he describes them. The crash of the Hindenburg really took place, there really was a big game hunter named Frank Buck, and Chaplin did, of course, make a film in which he played Hitler. Such facts are strung together in a lively narration and no doubt many readers learn from it something about the Depression era.

Because the narrator is a child he need not deal at any length with politics or the profoundly political artifact of the world's fair itself. Instead, the book is gently ironic, bordering on nostalgia. *World's Fair* is a comic work of domestic life, animated by the long-standing conflict between mother and father. Edgar early notes, "The conflict between my parents was probably the major chronic circumstance of my life. They were never at peace. They were a marriage of two irreducibly opposed natures. Their differences created a kind of magnetic field for me in which I swung this way or that according to the direction of the current."[18] The narrator cannot reconcile these opposed forces because he is a child. Only when he is old enough to do something independently and in secret—writing a contest-winning essay—can he bring about a temporary harmony in the family, as they celebrate his victory, first with an outing for ice-cream, and then with his prize, a free trip to the world's fair. Here his absent brother returns, the conflict between mother and father temporarily abates, and an ideal future world is substituted for their increasingly impoverished everyday lives.

Yet if the work's resolution is comic, the author does not adopt the organic mode of argument which has the greatest affinity for the comic mode. Rather it is "formist." As White puts it, the formist mode focuses on "The variety, color, and vividness of the historical field," and does not propound any general laws. Instead, it emphasizes "the uniqueness of the different agents, agencies, and acts which make up the 'events' to be explained."[19] Such a mode of argument is almost dictated by the use of a child narrator, who is virtually powerless to shape events until the end of the story, and who must retreat into private epiphanies and sufferings, studying his brother and his friends, listening to his favorite radio shows, playing with a girl friend, having appendicitis, or being beat up because he is Jewish. These disparate experiences hardly suggest a larger design, and the book unfolds rather slowly, with no strong narrative impulse.

III

Edgar displays an almost total recall of the past, wavering between the present and past tense, at times almost a child's voice, more often registering the memories of an older man, whose age and historical location are never clear. This voice seldom makes what it describes seem entirely immediate; what it recalls is often not fully realized in dramatic action. By comparing a similar scene described by Huck Finn and Edgar we can see the limitations of this kind of voice. Observe the contrast between two trips to the circus. As Huck describes it:

> And by-and-by a drunk man tried to get into the ring—said he wanted to ride; said he could ride as well as anybody that ever was. They argued and tried to keep him out, but he wouldn't listen, and the whole show came to a standstill. Then the people began to holler at him and make fun of him, and that made him mad, and he begun to rip and tear; so that stirred up the people, and alot of men begun to pile down off the benches and swarm towards the ring, saying, 'Knock him down! Throw him out!' and one or two women begun to scream. So, then, the ringmaster he made a little speech, and said he hoped there wouldn't be no disturbance, and if the man would promise he wouldn't make no more trouble, he would let him ride, if he thought he could stay on the horse. So everybody laughed and said all right, and the man got on. The minute he was on, the horse began to rip and tear and jump and cavort around, with two circus men hanging onto his bridle trying to hold him, and the drunk man hanging on to his neck, and his heels flying in the air every jump, and the whole crowd of people standing up and shouting and laughing till the tears rolled down. And at last, sure enough, all the circus men could do, the horse broke loose, and away he went like the very nation, round and round the ring, with that sot laying down on him and hanging to his neck, with first one leg hanging most to the ground on one side, and then t'other one on t'other side, and the people just crazy. It warn't funny to me, though; I was all of a tremble to see his danger. But pretty soon he struggled up astraddle and grabbed the bridle a-reeling this way and that; and the next minute he sprung up and dropped the bridle and stood! and the horse agoing like a house afire too. He just stood up there, a-sailing around as easy and comfortable as if he warn't ever drunk in his life—and then he begun to pull off his clothes and sling them. He shed them so thick they kind of clogged the air, and all together he shed seventeen suits. And then, there he was, slim and handsome, and dressed the gaudiest and prettiest you ever saw, and he lit into that horse with his whip and made him fairly hum—and finally skipped off to the dressing room, and everybody just a-howling with pleasure and astonishment.[20]

Edgar tells a similar story:

> It interested me particularly that in the circus there was one wistful clown who climbed the high wire after the experts were done, and scared himself

and us with his uproariously funny, incredibly maladroit moves up there. Slipping and sliding about, losing his hat, his floppy shoes, and holding on to the the wire for dear life, he was actually doing stunts far more diffi-cult than any that had gone on before. This was confirmed, invariably, as he doffed his clown garments one by one and emerged from the woeful little potbellied misfit as the star who headlined the high-wire act. In his tights and glistening bare torso he pulled off his bulbous nose and stood spotlighted on the platform with one arm raised to receive our wildest applause for having led us through our laughter, our fear, to simple awe."[21]

In each case, the boy observes an artist who disguises himself and pretends to be incompetent, but Edgar's voice lacks Huck's vividness and energy. He tells us his clown was "uproariously funny." Huck's language shows us that the trick rider is funny. Edgar's style is formal and "writerly", compared to Huck's vernacular, oral account. Huck describes a tumultuous scene of conflict, danger, hoax, and triumph. He sees everything, but he does not fully understand what he relates. He is not like the adult spectators who laugh at the tramp being thrown around by a horse. Rather, he is "all of a tremble to see his danger." As in so many other passages in that novel, his compassion exposes the inhumanity of others. In contrast, Edgar's clown scares the crowd as a whole, which has a unanimous sympathetic response to his danger. In fact, Edgar even says that the clown "scared himself." Thus Doctorow conflates the experience of the clown, the rest of the audi-ence, and his narrator, uniting them in a common response. In contrast, Twain amplifies the meanings and interpretations of events; the clown, the ringmaster, the crowd, and the narrator each give different meanings to the events, and all their responses are accessible to us through Huck's narration. Such a passage forces the reader to make an active interpreta-tion, just as in the famous passage where Huck decides to go to hell. Twain exploits differences and challenges the reader's morality, asking us to think about the audience that laughs at another's danger, or, elsewhere in the novel, to think about the relationships between whites and blacks or the psychology of the lynch mob. But Doctorow's narrator only passes lightly over most events, and all too many passages are like that on the circus. Often, all differences—between the adult and childish elements of the narrator, between audience and performer, and between reader and narrator—have been abolished. The abolition of difference simplifies the texture of the prose and solidifies the past into a series of static memories.

Such writing is not without its compensations, of course. It invites the reader to recollect personal experiences, and has as its chief virtue a kind of phenomenological evocativeness, bringing back the textures, smells, and sights of childhood. But in order to stand so close to the reader and perform this evocative work, Edgar's voice must have a mono-dimensional quality. He is not like Huck Finn, who admires the plaster apples and broken clock in the Grangerford's house, giving the reader sufficient information to reach precisely the opposite conclusion about their degree of refinement. Rather, Edgar sees the peeling paint of the Trylon and Perisphere at the fair while his parents do not.[22] He recollects the past wisely, and seldom allows himself to look foolish. Samuel Clemens found a creative play of differences in the invention of another voice, Mark Twain, who in turn invents the voice of Huck Finn, but Doctorow collapses the difference between himself, his narrator, and the reader. This strategy creates a voice that is meant to speak out of the past, of the past, and for the past, one that is meant to be the voice of the historical subject itself, telling both its own and the reader's story. In speaking for people dead and things mute, he becomes a ventriloquist, giving voice to the inanimate. This voice is not innocent like Huck's, yet it paradoxically provides much less social criticism. It is not a voice of experience, but rather one of what could be called "un-innocence." It sheds its historical knowledge of what came after 1940, and immerses itself in recollection. It is not a voice capable of extravagance, nor will it put us on guard by using dead-pan humor. Edgar's clown does not "clog the air" with seventeen suits, but rather doffs "his clown garments one by one." This tensionless voice seeks to animate all objects, yet is not animated itself. As one critic put it, "The on-going conflict between [the parents] Rose and Dave, which might have been crucial in the development of a real novel, generates surprisingly little tension; it is expertly recalled, but never fully dramatized. Much the same is true of young Edgar's fear of death and his growing awareness (and apprehension) of sex."[23] Such a voice only mentions family problems and social injustice without being energized. It is the reflective voice of the American middle class of the 1980s.

IV

This almost powerless child narrator with his dispassionate voice naturally can find no historical laws or forces to explain the larger pattern of events during the 1930s. At the same time, as Paul Levine observed before the work's appearance, "For Doctorow, the genius of American culture lies in its popular roots which flourish in the rich soil of ordinary life."[24] *World's Fair* testifies to this attraction to American popular culture, as opposed to imported European high culture. Edgar sees Babe Ruth do his radio show, goes to a double feature every Saturday afternoon, reads the funny papers, and chews Fleer's Double Bubble Gum. These evocations of popular culture charmed many reviewers. But what stance does Doctorow take toward popular culture as a whole? His earlier work shows that he is aware of its political implications. Daniel explains in *The Book of Daniel* that, "what Disneyland proposes is a technique of abbreviated shorthand culture for the masses, a mindless thrill, like an electric shock, that insists at the same time on the recipient's rich psychic relation to his country's history and language and literature. In a forthcoming time of highly governed masses in an over-populated world, this technique may be extremely useful both as a substitute for education and, eventually, as a substitute for experience."[25] Such sentiments closely resemble those of Horkheimer and others of the Frankfurt School, and they are appropriate for Daniel, since he tends to identify himself with the old left. Doctorow has also shown appreciation for Walter Benjamin in his essay "False Documents"[26] While these ideas are now regarded as outmoded by Marxists such as Fredric Jameson and by the reader-response school, they at least constitute a critical position from which to confront mass culture. They might have animated the depiction of 1930s popular culture, so that its chewing gum, movies, and radio shows could have become part of some larger pattern. At the least, the theories of the Frankfurt School might have informed the approach to the world's fair which gives the book its name, and whose description occupies more than thirty pages.[27] Not that Edgar himself should have engaged in dialectical analysis, but rather that, like Huck, he might have "unconsciously" revealed contradictions between values and behavior.

The title *World's Fair* leads one to expect that the novel will give a great deal of attention to the New York World's Fair of 1939, one of the most widely attended events in the first half of the twentieth century, with more than 45 million visitors. Compared with previous expositions, this one was more commercial, allowing corporations to have their own pavilions, rather than assembling similar activities in common halls for machinery, agriculture, electricity, the arts, and so forth. And because of the world-wide depression, corporate exhibits were better financed than most national pavilions. Thus the fair was not only a site where one might catch a glimpse of the future, it was a corporate vision of utopia. A novelist who deals with the fair as a part of history and as the culmination of a book permeated by popular culture has a duty to make such fundamental things clear. Indeed, Doctorow had deployed the technical means necessary for a full exploration of the fair's meaning: a clever young narrator, supplemented by multiple points of view from adults who add their "oral histories." But like any history this one selects and edits the materials in accord with its mode of emplotment; its use of the comic and formist modes makes a probing analysis difficult. Instead, Edgar visits the fair twice in accord with two related developments in the plot, each of which requires resolution, and the analysis of the fair is subordinated to these developments. The first visit culminates in Edgar's initiation into the world of sex, which transpires in the amusement park as he watches Meg's mother and several other women take part in an erotic attraction, "Oscar the Amorous Octopus."[28] The second visit brings the family together again in temporary harmony. On each occasion Edgar plays a more adult role than he could earlier in the novel, and these scenes in "the future" mark the end of his childhood.

The two visits may seem to provide a fitting fictional conclusion, but given Doctorow's insistence upon being considered on an equal footing with historians, quite other criteria must be applied. Edgar's "World of Tomorrow" reads much like descriptions in the fair's *Official Guide Book*. One comes away from the two works with a similar impression of the event. In each case the political and economic processes which brought the fair into existence are simply omitted, and the event seems to be a natural part of the time period, rather than the creation of certain interest groups. Edgar's father might have given a little declamation on this subject as he does on many others, but he regards the fair primarily as a possible source of new business in his record store, and he confines his criticism to the contents of the Westinghouse Company's "Time Capsule," and one mild comment on Futurama.[29] We need more information about the fair than Edgar is an a position to provide.

When Americans were mired in the long Depression, who but an elite could conceive and plan a $155 million project? World's fairs almost always lose money, yet $42 million in bonds were sold to finance a World's Fair Corporation. Buyers included the most powerful corporate leaders of the nation and Manhattan's social elite. State and local governments contributed $32 million in infrastructural improvements and services; foreign governments spent $30 million; and corporate exhibitors invested just over $50 million.[30] The common people played no part in fair planning, as they were not invited to invent appropriate themes, submit slogans, or buy shares in the fair. Nor did the New Deal agencies have an important role in creating the fair, although the Works Progress Administration (WPA) did have an exhibit. The fair was not a microcosm

Figure 5: The Consolidated Edison Exhibit *(Danish Royal Library)*

of 1930s America, but a corporate vision of what the future could be. Much is missing from the era of the New Deal, and the fair is at best a problematic symbol of American culture in the 1930s. Its themes, architecture, and organization lay entirely in the hands of an economic and cultural elite.

Yet *World's Fair* the novel gives virtually no hint that the fair was a severely edited expression of 1930s culture, and by any historian's standard is an inadequate account of the fair. It had two major themes, but one, the Presidency of George Washington, received virtually no attention in the novel.[31] By omitting virtually all reference to Washington and his times Doctorow ignores the way in which corporate leaders used patriotic

sentiments to bolster their social position and to provide a dramatic coun-
terpoint to the second theme: how science and technology would shape
the future. The novel emphasizes only this theme, which was expressed in
the slogan: "Building the World of Tomorrow with the Tools of Today."
Among the most popular exhibits were the Perisphere, Consolidated
Edison's "City of Light," General Electric's Stienmetz Hall, the Chrysler
Transportation Exhibit, and Futurama. Edgar visits all but one of these,
appreciating them rather uncritically. In viewing them he uses far less
intellectual energy than in his earlier analysis of popular song lyrics,
although he accurately notes how the fair's industrial designers played with
scale, "Everywhere at the world's fair the world was reduced to tiny size by
the cunning and ingenuity of builders and engineers. And then things
loomed up that were larger than they ought to have been."[32] But while he
is analytical enough to see that designers have continually played with the
scale of objects, Edgar does not criticize the strategy of representation as
a whole, which had the effect of eliminating the individual, who is either
miniaturized or enlarged out of existence. Such representations effectively
erased human beings as visible parts of the future, emphasizing instead
new machines, futuristic cars, idealized landscapes, and streamlined buildings. This realm of pure property was an ideal structure to house Edgar's
imagination, which seeks to minimize human differences and to recollect
sensory experience.

What does it mean when the citizen of the future has been made into
a stock figure in a model or amputated and blown up to be a representative eyeball or an enormous ear? In "The American Scholar" Emerson
noted how specialization had transformed the citizenry into grotesque
exaggerations. "The state of society is one in which the members have
suffered amputation from the trunk, and strut about so many walking
monsters—a good finger, a neck, a stomach, an elbow, but never a man."[33]
Where Emerson spoke metaphorically, in "The World of Tomorrow" the
individual was literally dismembered, and the parts of the body displayed
as objects of scientific understanding. Even more striking, of course, is the
radical difference between Emerson's emphasis on Nature and a world's
fair's characteristic emphasis upon the artificial. As anthropologist Burton
Benedict has noted, fairs are walled-off compounds, idealized consumer
cities that promulgate a future where everything is man-made, including
large replicas of natural objects. "At world's fairs man is totally in control
and synthetic nature is preferred to the real thing."[34]

At a time when the nation was in the throes of the Great Depression,
the world's fair was a shapely actualization of order. These displays could

"take people out of their ordinary routines and thus remove them temporarily from their usual positions in the social structure."[35] Once they were immersed in a fantastic environment, visitors such as Edgar and his family were prepared to receive new ideas about the larger direction of American culture. While attendance at a world's fair was open to all, the expense and considerable time fair-going required made it inaccessible to most working-class people. Those who traveled any distance to reach them were overwhelmingly middle- and upper-middle class. For them, the 1939 fair was one of the great events for a whole generation, providing a blueprint for the transformation of their own communities, which could only seem drab when they returned home.

At the most popular exhibits future marvels were suggested through dioramas and detailed models, and a smoothly controlled or automated presentation induced passivity in the viewer, who was urged to accept an ideal landscape as the probable future. The fairgoer's social construction of reality was transformed, because the objects displayed were no longer merely new consumer goods, they were talismanic parts of an ideal future. They proved that a benign future was already emerging into being, subordinating the rest of the fair to corporate exhibits.[36]

Edgar, like many of the fairgoers of the time, finds that the miniaturization of Futurama, the City of Light, and the Perisphere provides the psychological illusion of glimpsing another life from an Olympian height. Such representations effectively eliminated human beings as visible parts of the future, replacing them with a vaguely sketched everyman, not unlike the narrator whom Doctorow has merged with his times. Both the fair and the novel employ a representational strategy that permits each person to imaginatively enter a scene without impediment, but they do so by abolishing human difference. In effect, the novel uses Edgar's voice to provide a guided tour of the past similar to General Motor's tour of the future. For Emerson, the scholar had to become the seer for the rest of society, serving as its "delegated intellect. In the right state he is Man Thinking. In the degenerate state, when the victim of society, he tends to become a mere thinker, or still worse, the parrot of other men's thinking." In the Transcendentalist view, a child was less likely to be such a parrot, and as Twain later showed, it was possible to create a child narrator who made biting social criticism. But the voice of Edgar is not suited to that role. However successfully he may evoke the experiences of a child in the 1930s, at the fair he becomes a parrot of Norman Bel Geddes, the promoter and designer of Futurama. Indeed, *World's Fair*, the novel, has a rather curious affinity for the exposition's futurist landscapes. Like the fairground the whole design is harmonious, clean, uncluttered, efficient, and idealized. In each the unpleasant is either omitted, mediated by technology, or miniaturized to the point of invisibility.

At the end of the Futurama tour, the visitor emerged into the landscape of the fairground as though it was the future. Edgar says that "the amazing thing was that at the end you saw a particular model street intersection and the show was over, and with your I HAVE SEEN THE FUTURE button in your hand you came out into the sun and you were standing on precisely the corner you had just seen, and the future was right where you were standing and what was small had become big. The scale had enlarged and you were no longer looking down at it, but standing in it, on this corner of the future, right here in the world's fair!"[37]

V

The reader is no doubt meant to exclaim, "I HAVE SEEN THE PAST," as indeed many reviewers did. Novelist Anne Tyler declared in *The Detroit News* that "*World's Fair* is better than a time capsule; it's an actual slice of a long-ago world, and we emerge from it as dazed as those visitors standing on the corner of the future." Yet the reader of *World's Fair* has not seen the past. Despite its allegiance to chronology and verifiability, the book is less a depiction of the time period than an expression of the 1980s. Despite its brief references to poverty, anti-semitism, and racism, *World's Fair* is a curiously tensionless book about the Depression era. Only Edgar's unsuccessful father expresses radical opinions, and few of the other characters debate with him. Social problems are filtered through the newspaper and radio, they are not part of the action of the book itself. In its daily life there are no blacks, no strikes, no gangsters, no unionization drives, no migrants from the Dust Bowl, no Scottsboro Boys, no Indians being driven off their reservations, no public relations campaigns by large corporations, no hobos, no majority for neutrality, no Father Couglin, no revolutionaries. Edgar's father briefly refers to a few of these matters, yet they do not impinge directly upon the action of the book. Mediated by the radio, newspapers, and a child's consciousness, history does not hurt, and the 1930s being remembered is appropriate for the middle class of the 1980s. It embraces popular culture, which has also been fashionable for the last two decades, and it finds the strongest kind of community not in ethnicity or race but in fragile, constantly shifting allegiances to radio programs and consumer goods, binding people into what Daniel Boorstin has termed "consumption communities." For Boorstin, as for Edgar, "The modern American, then, was tied, if only by the thinnest of threads and by the most volatile, switchable loyalties, to thousands of other Americans in nearly everything he ate or drank or drove or read or used. Old-fashioned political and religious communities now became only two among many new, once unimagined fellowships. Americans were increasingly held to others not by a few iron bonds, but by countless gossamer webs knitting together the trivia of their lives."[38]

The book's mode of ideological implication seems clearly liberal, in White's meaning of the term, because it imagines an improved world in the future, achieved neither through any violent change nor through the operation of any historical laws. Yet this liberal encodation is itself undercut by the historical position of the reader, who, unlike the narrator, sees the 1930s from across a chasm of war, holocaust, and social unrest. Thus the reader may well choose to encode the book in an anarchist mode, seeing Edgar's childhood as a remote past of innocence from which men today have fallen into a more corrupt social state. Cushing Strout's conclusions about *Ragtime* illuminate the weaknesses of *World's Fair* as well: ". . . one cannot read *Ragtime* consistently either as playful fantasy or as serious history. It is too historical for farce, too light-hearted for the rage of black humor, and too caricatured for history. . . . *Ragtime* appears to struggle with the complexities of history, but is only a clever trick. Modern art rightly celebrates artistic freedom, yet its victories are cheapened when there is no resistance from the pressure of a responsibility to historical consciousness."[39] *World's Fair* is more conscientious than *Ragtime* in handling individual historical facts, but it also plays fast and loose with genres, escaping from both the conflicts of history and the self-revelation of autobiography. Its narrator is wise beyond his years when criticizing song lyrics but retreats into childishness rather than deal with the harder questions of poverty, social class, corporate hegemony, and politics.

The book is too autobiographical to be considered a history of the period and too historical to be taken as a work of fiction. It reveals little that was unique in the life of its author, and it simplifies the 1930s into a series of harmless vignettes. Crossing the boundary between history and literature requires more than mere veracity. As White reminds us, the energy and interest of all great historians arises not merely because they are accurate. The story they tell must deal convincingly with a broad range of facts, and offer a new way to see and organize the materials of the past. Doctorow knew this when he wrote *The Book of Daniel*. But *World's Fair* does not probe a historical question or provide a new way to see the 1930s. Its narrators do not bear witness to a world as complex or as morally ambiguous as Huckleberry Finn's. Instead, the novel urges us to accept the limited moral vision of a child without granting us youth's compensating verbal energy or its all-encompassing eye. It asks us to accept evocations of popular culture uncritically and to believe in the corporate public relations of 1939. It asks us to accept the form of oral history without the probing questions of an interviewer. Such a book is at the farthest remove from the post-structuralist project of demystifying narrative, which emphasizes not unanimity but contradiction and the creative play of difference. One hopes that in the future Doctorow's enviable clarity and verbal facility will again be harnessed to adult narrators like Daniel, and that they will encounter more stubborn resistance from historical facts. After declaring that there is no history or literature but only narrative, Doctorow has opted for a veracity of the sort associated with Old Sturbridge Village or Henry Ford's supermarket of history, Greenfield Village. The short, smoothly flowing sentences he has crafted are like the gleaming, polished surfaces of these reconstructed museums. They are closer to the streamlining of Futurama than to the gritty complexities of the 1930s.

III

NARRATIVES IN SPACE

SEVEN
Electrifying Expositions, 1880–1939

As the preceding analyses of *The Great Gatsby* and *World's Fair* demonstrate, narratives contain an implicit relationship to the technologies of their time. Where Nick Carraway's story expressed the contradictions that electrification introduced, Doctorow's child narrator effaced similar complexities in the 1930s. If there were four competing narratives for rural electrification, surely there were as many ways to look at the 1939 World's Fair. The literary work is articulated in relation to the cultural ground from which it emerges. That this ground is constantly shifting can be seen by looking at the history of expositions themselves. While each fair seeks to appear unique and focuses on the future, like a historical narrative its difference is constructed through contrast with preceding expositions. By its very nature the world's fair must constantly seek innovations, which almost always come in the form of new technologies. And of all the technologies employed to dazzle fairgoers, electric lighting was one of the most important. Spectacular illumination made it possible to display two landscapes in one space, one by day and another by night. This innovation was constantly improved upon between 1880 and 1939, and as techniques became more sophisticated, designers were able to combine lighting with an ensemble of other electrical technologies to create narratives of abundance and progress.

I

A visitor to the Musée d'Art Moderne de la Ville de Paris cannot avoid seeing a painting depicting the history of electrification, by Raoul Dufy, because it covers 600 square meters. "La Fée Electricité" was commissioned by the Paris Electricity Company (la Compagnie Parisienne d'Electricité) for the 1937 Exposition Universelle. A historical panorama, it depicts 108 recognizable figures in the history of electrification, most of whom are on the lower panels at eye level, as well as suggestive glimpses and fragments of the electrified landscape, including advertising signs, power stations, and transmission lines. The overall color scheme is bright, predominantly greens, yellows and blues, with splashes of red. Given the sponsor, many of the figures chosen were French, but Dufy also included important figures from Russia, Germany, England, America, and other nations.[1] It is not my purpose to analyze or even summarize this enormous work. But note that it embodies a traditional sense of history that emphasizes individual biography. Electrification is the story of great men, whose inventions changed the world. It is not the story of corporations and their markets, nor is it a story where the public plays an active role. The history of the great expositions themselves presents quite another narrative, however, in which large corporate sponsors played the major role in shaping expositions, as they catered to the public.

Yet none can deny that the inventor and the scientist were important figures at international expositions, almost from their inception in 1851. One could construct a history of the ways inventions were used to represent the "universal" culture of science, to show the superiority and the progress of Western culture, and to celebrate individual genius. Visits to the fairground by famous inventors and scientists were often made into grand occasions, and at times they were prominent in the official openings of fairs. To take but two examples, George Corliss was a central figure at the inauguration of the Philadelphia Exposition of 1876, and Albert Einstein spoke briefly at the opening of the New York World's Fair of 1939. Yet over time exhibitions passed through a series of four distinct stages. The emphasis on individuals and particular machines that Dufy depicted in his painting was the hallmark of the first stage, from 1851 until c. 1875. After then, although individuals and their creations continued to be prominent, they increasingly gave way to a systems approach, in which ensembles of objects were ordered by a dominant technology, at first the steam engine and then electrical systems. This second stage, which focused on interior displays, gave way at the end of the century to sophisticated efforts to unify entire fairgrounds as coherent electrified landscapes. This third stage, characterized by unity of theme, lighting, and overall design, was fully achieved by 1901 at the Pan-American Exposition in Buffalo. In the fourth stage, during the 1930s, planners perfected electrical control of the exhibits themselves, culminating in the multi-media simulations of the future that proved so popular with the public at New York's 1939 fair. The four stages of electrification at expositions over-lapped, of course, and each persisted as a part of the ensemble of techniques. The order in the sequence was not accidental, as it clearly recorded changes in technology. Most obvious was the increasing sophistication of lighting technologies that moved from the powerful glare of arc lights (1878) to the more subtle and controlled effects of incandescents (1881), to the special effects of floodlighting (after 1907), to the smooth lines and precision of neon (after 1920), incorporation of lighting into architectural elements (late 1920s), and finally fluorescent tubes (after 1937). Equally impressive developments occurred in many areas, including the transmission of sound over loudspeakers, film projection, the automatic movement of objects, and the sequencing of special effects.

Early electrical displays explained new products and early fairs featured inventors. The stories they told were much like that implied by Dufy's sequence of inventors. But as technological systems improved, hagiography gave way to pedagogy, as exhibits sought to educate the public about electricity, even as it surged out across the fairground. As electricity became increasingly common, however, education gave way in turn to lively demonstrations of what could be done with it. Rather than emphasize the scientific nature of each advance, exhibits focused on powerful special effects. This general shift from education to entertainment was accompanied by more powerful technologies of representation, which permitted far more dramatic effects.[2] Landscape and narrative fused in these three-dimensional displays.

II

During the first three decades of world's fairs steam was the dominant technology. Steam engines were familiar to Victorians, who had no difficulty understanding their general principles or their systems of drive-shafts, gears, belts, valves, pulleys, cut-offs, and levers. The creation and transmission of power was a visible process that could be traced with the eye and the hand. It had come to seem a part of common sense. Early fairs celebrated the age of steam by erecting very large engines, which the public judged by their size and smoothness in operation. The apotheosis of this trend was no doubt the erection of the giant Corliss Engine at the Philadelphia Exposition of 1876. It drove all the devices in Machinery Hall, and it quickly became the central symbol of the fair, not least because its enormous fly-wheel moved just slowly enough so that the eye could follow its rapid, endless, virtually silent revolutions.[3]

In contrast, the electrical systems that began to be displayed at fairs during the next fifteen years worked according to principles few could grasp. Direct current moved mysteriously over wires or along trolley rails to provide light and power. Between 1876 and 1890 appeared the wonders of the telephone, phonograph, practical electric light, loudspeaker, and electric railway. Each was at first displayed individually, with an emphasis on the inventor, as was the case at the 1881 Exposition Internationale d'Electricité in Paris.[4] France had the lion's share of exhibits, 55 percent, but there was also significant participation from Germany, Great Britain, Belgium, and the United States. For the first time visitors could travel on an electric train, built by Siemens, which took them 500 meters, from the entrance to the main exhibit hall.[5] This structure contained 10,000 square meters, organized by nationality and biography. Two central figures were Thomas Edison and Alexander Graham Bell, whose electric lighting and telephone systems were given prominent positions on the main floor. Two entire rooms were dedicated to Edison, featuring not just the light bulb itself but his newly developed system with its wiring, sockets, fuses, generators, and the other necessities of distribution. Bell displayed not just a working telephone but a rudimentary telephone exchange. Visitors left their exhibits with exciting intimations of what an electrified world might be like, but they had only seen it in fragments. In the subsequent three stages, world's fairs abandoned the biographical approach to display, favoring the deployment of entire systems.

During all four stages, it must be stressed, electrical displays served economic interests, first those of entrepreneurs such as Siemens, Edison, and Bell, and later the interests of corporations they founded. At every fair from 1881 until 1939 an entire building or more was devoted to electricity, and these structures were among the largest, most central, and most popular halls at expositions in these years. Here the impulse to celebrate inventors naturally survived longest. For example, at Chicago in 1894 the most prominent object in the Electrical Building was a "Tower of Light" covered with ten thousand lamps, surmounted by a "mammoth incandescent lamp built up of about 30,000 cut-glass prisms. . . .By means of a commutating device within the base of the tower, the lights were thrown on and off in a great variety of charming combinations."[6] It honored Edison and drew attention to an educational display that explained how electric lights were manufactured. Yet such displays gradually took second place to spectacles staged for their own sake, as fair organizers began to use electrical technologies in the overall design of the fair. In the 1880s electric lighting moved from isolated demonstrations to being a fair's preferred form of illumination, and by the 1890s electric traction had become the preferred form of exposition transportation. Electrical corporations became deeply involved in overall planning of every world's fair, gathering new ideas at annual electrical trade fairs where engineers developed the expertise needed to display products and systems at the major expositions.

Electrification quickly became more than a mere expedient that permitted the fairgrounds to remain open in the evening. It was an ideal element of display, at once refined, abstract, and modern. It was an elusive concretization of abstract qualities that could be known only by its effects: light, heat, and power. Spectacular lighting was dramatic, non-utilitarian, and universalizing. Its introduction did not detract from other exhibits, but rather provided a brilliant canopy, connecting diverse elements in one stunning design. Little wonder that many fairs made illuminated towers their symbols, including the Eiffel Tower, Buffalo's Electric Tower, San Francisco's Tower of Jewels, and New York's Trylon. Electricity became more than just the theme for a major exhibit; it provided a visible correlative for the ideology of progress and abundance through technology.

Yet granting that electricity had many advantages as an element of display, its extensive use was hardly disinterested. Edison and his main American rivals, George Westinghouse and Thomson-Houston, all moved quickly into foreign markets. Many of Edison's skilled workmen involved in the development of the electric light system were immigrants, and they returned to Europe in the 1880s, to start utilities and electrical-equipment manufacturers. Francis Jehl returned to Austria–Hungary, Thomas B. Thrige went home to Denmark, and S. J. Bergmann went back to Germany, where he established a company that eventually employed 30,000 men.[7] Thomson-Houston sent agents to Europe as well, bringing its international holdings in 1892 to the formation of General Electric. Westinghouse early entered the British market, and the companies they spun off remained two of the three largest electrical corporations there in 1930.[8] Often the inauguration of new companies and participation in expositions were simultaneous, and these efforts intertwined. Edward Johnson promoted the Edison telephone and lighting systems in England, starting in 1878. He was also responsible for his company's first central station in London, which opened in 1881, and for the Edison exhibit at the London Crystal Palace Exhibition of 1882. Every subsequent British exposition was influenced not only by the example of American expositions but by American electrical corporations.

Similar patterns of influence can be found on the Continent. International General Electric was the key member of an international cartel of manufacturers, and it held a sizable percentage of the stock in European street traction businesses and in lamp-producing firms. In 1929 General Electric had substantial daughter companies in Austria, Belgium, France, Great Britain, Germany, Greece, Holland, Hungary, Italy, Portugal, Spain, and Switzerland—in short everywhere except Scandinavia.[9] By 1931, 25 percent of General Electric's total equity was held in Europe, concentrated in Germany (9 percent), England (8 percent), France (4 percent), and Holland (3 percent).[10] In France, where it had invested only half as much as in England or Germany, GE subsidiaries built most of the traction systems in Paris, Le Havre, Lyons, Marseilles, and smaller cities. By 1922 one subsidiary, Cie Française Thomson-Houston, employed 9,500 people in nine factories, including a plant making telephone equipment, mining equipment, electric lights, and some domestic appliances. The Compagnie des Lampes, formed in the merger of several existing companies, including the lighting interests of Cie Française Thomson-Houston, was the largest producer of incandescent lamps in France.[11] General Electric also had opened a new research laboratory in France in 1922. American electrical companies dominated many European markets, and often had seats on the boards of ostensibly competing firms.[12] Given these interlocking financial interests and the web of personal ties that connected the pioneering entrepreneurs, there was a rapid interchange of electrical display techniques at world's fairs between the two continents. Symptomatically, the first electric sign was produced by an American (Edison) but it was first displayed in Europe, at the Paris exposition of 1881.

III

By the middle of the 1880s exhibitions were entering the second stage, when electrified interiors drew vast crowds to events that previously would have closed at dusk. At the Louisville Southern States Exposition of 1883 the Edison Electric Lighting Company installed 4,600 light bulbs in the exhibition interiors, creating a sensation.[13] Electrical special effects were further perfected the following year at the International Electrical Exhibition in Philadelphia, which featured an indoor fountain thirty feet in diameter, whose fifteen jets were periodically made the center of attention by extinguishing all other illuminations and training colored lights on the streams of water. Equally striking was the Edison exhibit, a thirty-foot column covered with more than one thousand lights that "climbed" around it.[14] A reporter noted that "the crude buildings hurriedly erected without any attempt at finish for a temporary purpose, were transformed into a temple of light, which at the first glimpse evoked expressions of delight from every beholder."[15]

In these early years, when special effects were still experimental and electrical equipment still delicate, most displays remained inside the buildings, while arc lights illuminated the grounds. In 1888 at the Centennial Exposition of the Ohio Valley and Central States, one hundred powerful arc lamps were used outside the exhibition buildings. Inside, one could see the delicate, colored incandescents.[16] By this time in some of the exhibits lighting had already moved away from being a matter of scientific demonstration and become a form of entertainment. There were faint light bulbs in the eyes of carved animals and inside Japanese paper fish, for example, and some attendants in the entry hall wore "helmets surmounted by a powerful electric lamp" that blinked on whenever an attendant stepped on a steel plate. The most impressive effect, again, was an electric fountain. All of these were startling novelties, still presented as isolated displays.

These techniques were further developed at the Paris Exposition of 1889, where "a nightly show of illuminated fountains entranced crowds with a spectacle of falling rainbows, cascading jewels, and flaming liquids, while spotlights placed on the top of the Eiffel Tower swept the darkening sky as the lights of the city were being turned on."[17] Other special effects were created by "outlining the prominent architectural features of the building facades with light." This technique was an "amplification of effects previously obtained with gas jets," and it was a former expert in gas illumination, Luther Stieringer, who carried this system of exposition lighting to its highest point in a series of American fairs in the decade from 1894 until 1904.[18]

By the Columbian Exposition the centrality of electric lighting at fairs was firmly established, and the Chicago fairground installed more lighting than any city in the country. Many visitors saw more artificial light in a single night there than they had previously seen in their entire lives. Earlier effects had been further developed. Notably, mobile jets of the outdoor electric fountains at either end of the Court of Honor shot water high into the air and wove complex patterns against the night sky. Spotlights underneath the fountains were fitted with colored filters, which permitted operators to create symphonies of color as they spewed 44 thousand gallons of water a minute in kaleidoscopic variations, to the accompaniment of band music.[19] Photographic technology of the day could not capture these displays, but by all accounts they were among the most popular events. One observer noted that long before six in the evening crowds began to gather around the Court of Honor, ". . . eager to secure a good position from which to behold the illuminations." But the spectacle covered "so wide a range of territory, that it is no easy matter to obtain a position where all can be surveyed. The electric fountains and Administration Building in a blaze of glory are at the west end; the magnificent pyrotechnic display is eastward of the lake; the surface of the grand basin is covered with floats from which shoot up numberless fiery serpents; all along the roofs of the Agricultural and Liberal Arts Buildings are lines of flickering flambeaux."[20] Many, including Henry Adams, preferred the panoramic view from the Ferris Wheel.[21] Theodore Dreiser preferred the view from an electric launch, where "a feeling of the true dreamlike beauty of it all came to me, at first only as a sense of intense elevation— not wonder, but elevation at being permitted to dwell in so Elysian a realm. . . .Then followed an abiding wonder."[22] Electrical illuminations charmed even the most critical observers. The novelty was not only the new landscape of light, but also the way it replaced the monumental buildings of the day. This stunning substitution was in itself a narrative of progress.

The lesson was not lost on the organizers of the 1898 Trans-Mississippi Exposition in Omaha, who hired Henry Rustin to design an entirely new style of illumination. Consulting with Stieringer in New York, he developed an outdoor system based on incandescent bulbs. As at Chicago, neoclassical buildings were sited along a canal 150 feet wide, but instead of the harsh glare of powerful arc lights thousands of incandescents festooned the buildings. At dusk on opening day a vast crowd filled the grounds to see the illuminations.

> Just as the outlines of the faraway buildings began to grow indistinct in the deepening shadows, a single cluster of electric lamps on each side of the lagoon was lighted. Then another and another until the row of pillars that circles midway between the lagoon and the buildings was crowned with incandescent luster. Another turn of the switchboard and the circle immediately surrounding the lagoon added its radiance and flashed golden bars across the water. In another instant the full circuit was opened and every outline and pinnacle of the big buildings blazed with light. The effect was indescribable. . . It was magnificent beyond comparison or comment and the immense crowd that had been waiting patiently for the moment gazed in dumb admiration. For a few seconds the vast court was as silent as though it was peopled with wax figures. The approbation of the people was vented in a volley of cheers and handclapping. On every side were heard the most extravagant expressions of admiration. . .[23]

Figure 6: Night View, East Half, Grand Court, Omaha Exposition, 1898 *(Omaha Public Library)*

Harper's Weekly found it "superb," and far more subtle and beautiful than an illumination its reporter had recently witnessed in Paris.[24]

Two years later the Paris Exposition adopted the techniques pioneered in Omaha and kept the fair open long into the night. *The Revue de Paris* marveled that a "simple touch of the finger on a lever" could transform the Monumental Gateway with "the brilliance of three thousand incandescent lights, which, under uncut gems of colored glass, become the sparking soul of enormous jewels."[25] The banks of the Seine and its bridges were also illuminated, giving Parisians a foretaste of the electrified urban landscape that soon would become "normal." The spectacle was a great popular success, and it masked to a considerable extent the fact that French engineering was falling behind Germany and the United States.[26]

Following the successes of Chicago, Omaha, and Paris, organizers of Buffalo's Pan-American Exposition of 1901 made electrification their central theme, in a fair that celebrated the completion of the first hydroelectric power stations at Niagara Falls. Rather than blinding visitors with powerful lighting, as Chicago's arc lights had, Buffalo's engineers greatly expanded upon the Omaha model. At the center of the fairground stood a 400-foot Electric Tower, covered with 40,000 small bulbs, and surrounded by a Great Court with 200,000 more. These made the whole scene appear like a three-dimensional impressionist painting. As in Chicago and Omaha, there was a ritual at dusk, when all lights were extinguished and waiting crowds watched a carefully orchestrated event. It began like daybreak, with a pink flush of dawn on the Electric Tower, which then deepened into red and shifted slowly to yellow, as the new "day" illuminated the fairground.[27] Buffalo was the first to use colored lights in a unified design, with the brightest colors and subtlest effects in the center at the Electrical Tower. The Louisiana Purchase Exposition of 1904 used similar techniques, making illumination a central part of the fair experience. The official publication celebrated the night view of the grounds, which was already a vision worth seeing from a mile or more away. "The distance not only lends enchantment to the view, but mellows the scene to a soft glow, soothing to the eyes. One beholds glowing though the darkness, long lines of little lights, broken here and there into fantastic designs." In this new aesthetic, which would soon be extended to American cities, electricity dematerialized the built environment.[28]

Electrical effects had passed far beyond the uncoordinated, special effects that started inside exhibition buildings and migrated outside by the end of the 1880s. Outdoor lighting, begun as a blatant glare of arc lights, had evolved into impressionistic colors and subtle patterns at Buffalo and

Figure 7: Night View, Court of the Four Seasons, San Francisco Exposition, 1915
(General Electric Photographic Archives)

St. Louis. These fairs were fully coordinated aesthetic designs, adding to the architectural unity imposed on previous exhibitions a comprehensive plan for illumination. Such plans next began to emphasize floodlighting, which was first seen at regional events such as the Hudson–Fulton Celebration of 1909 and had been fully developed by the 1915 San Francisco Panama–Pacific Exhibition.[29] There, as already described in the chapter on the electrification of the West, buildings were lighted by spotlights, displaying the fairground in sumptuous detail. Multi-faceted crystals embedded in the walls (called nova-gems) glittered and gave off a rainbow of iridescence. There was "no blaze or glare" for this was not white light but "softened and tinged with the warmest and mellowest of colors."[30] Where previous fairs had emphasized the sharp distinction between day and night, the Panama–Pacific sought to erase that distinction. The spotlights gradually came on at dusk mingling the natural and the artificial, and after the sun went down a bogus sunrise was staged in the west followed by an artificial aurora borealis that extended from the Golden Gate to Sausalito. Other stunning effects included "fireless fireworks," developed by GE technicians in the previous decade. To provide a medium to work in, a stationary mountain steam locomotive was driven at the equivalent of 60 miles per hour, with its brakes on, producing enormous clouds of steam and smoke. As *Scientific American* reported, the special effects were a "wonderful spectacle" with its "Fairy Feathers," "Devil's Fan," "Sun-burst," and "Plumes of Paradise."[31]

The net effect of electrical displays before 1915 was far more than to sell individual products. They created a dream landscape where all unsightly details could be excised. The utopian vision they expressed catered to the genteel and to the middle class; illuminations exalted neoclassical buildings and the horizontal city preferred by beaux-arts architects, in opposition to the emerging vertical city of the skyscraper. Exposition illuminations also helped to impose a progressive order on the world, as forms of electric light, heat, power, and transportation appeared to be part of a rapid evolution from savagery to civilization. Dreiser noted in 1894 that a sense of progress toward a higher ideal was implicit in the creation of a fair's order and harmony, in a place where less than a century before there had been no city at all.[32] Electrical displays gave the visitor an explanatory blueprint of the future, promising to ease work and ban ugliness. Yet in the first three stages of development this utopian blueprint remained an outline. It was not yet developed in detail.

IV

The hiatus in major fairs during the 1920s occurred during a period of rapid electrification in all areas of American society. By the time expositions were in vogue again during the 1930s, public illumination was widespread, the majority of American homes had electricity, and the night skyline was a fact of life. Spectacular lighting, central to the fairs in Chicago, Buffalo, and San Francisco was a commonplace, and it could no longer play as important a role. Instead, it became an element in a larger ensemble of techniques, including film, amplified sound, and precise orchestration of the fairgoers' experience.

All the world's fairs before World War I had emphasized the "white city" as the logical corollary of the beaux-arts style in architecture. In contrast, the lighting designs of the 1930s represented a compromise between the international style and the colorful vernacular of Times Square. Fairgrounds emerged as static designs. Instead of the discordant, flashing lights of the commercial zone with their illusions of movement and reiterated messages that actively competed for attention, the Chicago Fair of 1933 and the New York Fair of 1939 were virtually still patterns. Color schemes were introduced, so that each avenue or area had a dominant tone. The resulting landscape aestheticized the Great White Way, preserving its hues but not its surging energy.

The Chicago Century of Progress Exposition of 1933–34 experimented with windowless buildings, whose interiors were completely lighted and ventilated by artificial means. This artificiality was presented as a guarantee that light would be equally good at all times.[33] Outside, these windowless structures were painted in strong primary colors and lighted to create abstract patterns. Planners had much stronger bulbs and a wider range of colors available than in 1915. Furthermore, neon, whose use had previously been restricted by patents, was also now at their disposal. The basic approach, however, was much like that in San Francisco. Floodlighting served as the basic form of illumination, with newer kinds of lighting as accents. As in 1915, a battery of powerful searchlights were installed at one end of the fairground, but in Chicago they were far more powerful, almost 2 billion candle-power. Despite these similarities, the overall effect was far different. The architecture was no longer neoclassical, and primary colors were used rather than the pastels in vogue a generation earlier. The catalog declared that "As you mingle with the crowds at night, you stand in the greatest flood of colored light that any equal area, or any city of the world, has ever produced."[34] Attempts to impose a uniform color scheme on exhibitors were not very successful, however.

Extending these developments in Chicago, the New York Fair of 1939 banned floodlights on all buildings except for the central theme exhibit, the Perisphere. Instead, exterior lighting was built into structures as part of the architectural design. Rejecting floodlighting, which made buildings appear at night much as they would in the daytime, organizers decided that "each building" would have "a night appearance quite different from its daytime appearance." Realizing that people had learned how to see conventionally lighted structures, they wanted defamiliarization: "a luminous design of a static character, that by no possibility could be had by natural day light."[35] Exterior lighting emphasized paintings, plaques, sculptures, banners, fountains, and architectural details, and it often intentionally left walls or roofs in darkness, creating dramatic contrasts. In this way the fair at night not only looked far different than by day, but it became a textualized landscape that demanded to be read. This plan also promoted a line of new products that General Electric had brought on the market. Called "luminous architectural elements," these could be used inside and out to create new visual effects. Interiors could be lighted from "luminous recesses, coves, grooves, and coffers" of the sort used in the lobby of the Chrysler Building.[36] More important for exterior lighting were luminous panels, which offered subtler effects that floodlighting, because it was easier to control the distribution and the brightness of the light. Luminous pylons and columns were also sold after 1931, effectively providing a new visual vocabulary for expositions.[37] Another important lighting innovation of the late 1930s was fluorescent tube lighting, used first in Paris in 1937 and then more extensively in New York.

Yet special visual effects on the outside of the pavilions were far less important than the integrated electrified technologies used inside.[38] No longer were exhibitors content to present displays according to a traditional museum format, with cases and objects placed along a heavily captioned route that a visitor might or might not travel. Designers at the New York 1939 "World of Tomorrow" synthesized light, sound, animation, and miniaturization in exhibits that provided visitors with the powerful illusion of seeing the world of the future. This was achieved for example in the Perisphere theme center, where visitors looked down from a slowly revolving balcony at an idealized "Democracity." Consolidated Edison used a similar approach inside a block-long building that held a miniaturization of New York City. Here the public saw a simulation of one day pass by, while a recording told them how electricity kept the city functioning (see figure 5, p.105). The most popular exhibit of all, General Motors' Futurama, simulated the landscape of 1960 seen from an airplane. The twenty-minute flight covered hundreds of miles over a vast miniaturized world, a "continuous animated panorama of towns and cities, rivers and lakes, country and farm areas, industrial plants in operation, country clubs, forests, valleys, and snow capped mountains." It contained more than half a million visible houses and 50,000 scale model automobiles, many of them moving along superhighways. A narrator explained the scene and dilated on the scientific wonders of the future.[39] Such exhibits coordinated pre-recorded music and narration with sequences of lighting effects that played over ideal landscapes. The visitor was not allowed to proceed at his own pace, but rather was moved through automatically. The exhibitor controlled both the sequencing and the interpretation, creating a narrative that transported fairgoers into the future.[40]

In the 1880s an indoor electric fountain or an electric sign had been enough to transfix the visitor. No sequence of events was required. In the 1890s for a few minutes each night the spectacular lighting of an entire fairground held the multitude in its thrall, and each successive fair searched for ways to heighten and lengthen the drama. But fair designers found that because corporate headquarters and government buildings copied their spectacular effects the public became harder to impress each year. By the 1930s the appetite for novelty could no longer be satisfied with lighting effects alone. Instead, complete environments had to be created, simulacra of an ideal future, where human needs would be met by electrical machinery. At the 1939 "World of Tomorrow" electricity in all its forms was assumed to be part of every ideal landscape or industrial process. Corporations had been so successful at making expositions electrifying that by 1939 the marketing of this new technology was virtually complete. The electric landscape had become an assumed background to daily life. In a scant sixty years electricity had moved from the foreground of consciousness to a tacit assumption about the shape of the lifeworld.

The actors in this story of electrifying expositions were not Dufy's heroic inventors but American corporations, who continually improved their methods of display. They had used world's fairs as proving grounds to develop an aesthetic with wide appeal. Not only had they made electricity an assumed part of everyday life, they had learned that pedagogy was not what the world's fair public wanted. They had abandoned the museum format, as displays became increasingly theatrical, interactive, and programmed, transporting the fairgoers effortlessly through an artificial landscape. The most successful exhibits immersed people in an environment and told them one or another version of the narrative of abundance.

In contrast, most Europeans still regarded expositions as expressions of national cultures rather than as corporate opportunities. For the most part they retained a museum style that proved ill-suited to addressing the American public.

EIGHT

European Self-Representations
at the New York World's Fair of 1939

I

During the 1930s Western European governments realized that they would need the assistance of the United States if they were to survive a conflict with re-armed Germany. Yet in these years the United States was neutral and had a small military establishment. The vast majority of the citizenry wished to keep the nation neutral, and as late as March 1941 a Gallup Poll found that an overwhelming 83 percent of the American people opposed entering World War II.[1] This single statistic from after the end of the New York World's Fair of 1939 might seem to suggest that it made little impact on the popular view of foreign policy. But while Americans hoped to avoid fighting the war, by 1941 they had reached a consensus about which countries they supported, and they did so during the time the Fair was open. The 45 million people who paid admission to see "The World of Tomorrow" rather surprisingly preferred the international exhibits to the lures of the amusement zone,[2] and the most popular foreign exhibits belonged to America's future allies, Russia, Great Britain, and France. They had spent millions of dollars on their pavilions, recognizing that they needed to court American public opinion and create a favorable impression. This sense of a community of interest, which was at first confined to a minority, found a focus at the fairgrounds.

For more than a century the relationships between Europeans and Americans have been strikingly illustrated by the exhibits at the world's fairs, which have occurred at regular intervals since their inception at the London Crystal Palace Exposition of 1851.[3] Expositions are attractive for investigation because they offer both concrete expressions of ideology and a measure of public response to the various national displays. Like museums, they are official representations of culture, and like any representation, the national exhibit at a world's fair is mediated. There are at least three kinds of culture operative: the internal culture of the world's fair as an institution, the culture of the host nation, and the various cultures of the exhibitors. These three are by no means easily compatible, and the achievement of a successful world's fair or individual exhibit requires compromises and adjustments between conflicting claims.

First, there is the internal culture of the world's fair as an institution. It evolved from a series of national fairs in Europe into the Crystal Palace Exposition of 1851 in London. From then on no decade lacked several fairs, and the institution evolved an international network of experts who were skilled in the diplomacy, fund-raising, and staging of expositions.[4] No exposition was sui generis, as each drew upon the cumulative experience of previous fairs, often employing some of the same personnel.[5] When Cincinnati, Atlanta, New Orleans, or Omaha staged fairs in the nineteenth century, they did not start from scratch, but studied the expositions that had preceded them. They almost always sent delegations to the fairs immediately before their own. The New York Fair's organizers had ample opportunity to study expositions in the 1930s, most notably in Chicago (1933–34) and Paris (1937), but also in San Diego (1935–36), Dallas (1936), and Cleveland (1937).

One may well ask what particular values characterize exposition culture between 1851 and 1939. In general, during these years expositions value peace and banish reference to war. They embrace new technologies, trumpet the advance of Western civilization over the globe, and advocate the existence of commonalities between the United States and Europe. The logic of the world's fair also stresses the future much more than the past,[6] and the primacy of culture over nature, which exhibits either replicate or improve upon. World's fairs are almost entirely secular institutions. They seldom show a preference for any denomination within their gates, and usually all representations and practices of religion are confined to a single space of minor importance.[7] Expositions are also partial substitutes for religious shrines, offering a site for the pilgrims of modern society to express their faith in technology and progress. In short, exposition values do not simply mirror the world outside the perimeter of the fairgrounds. Indeed, they often depict the reversal of existing economic conditions, as was the case during the sudden burst of American fair activity during the 1930s.

The most pointed contradiction between the world depicted within expositions and actual conditions is the emphasis on peace and the reluctance to display military technologies. Even the most intense ideological rivalries, such as that between Nazi Germany and the Soviet Union at the 1937 exposition in Paris, primarily find expression in monumental architecture and displays. World War II broke out in 1939, and as a result, several national exhibits at the New York Fair were forced to close. Yet within the fairgrounds the war was seldom mentioned, and technology was presented almost exclusively as a benign force.[8] Likewise, most of the interwar expositions presented imperialism as a tale of progress and enlightenment, even as anti-colonial movements were developing around the globe. If the culture of expositions sought to impose its vision on the world, it was in fact dependent upon the international situation. Nevertheless, the fiction expressed within the grounds of an exposition usually is that it represents the world toward which all of civilization is evolving. Fairs usually emphasize the future, even if centennials and other anniversaries are the formal excuses for staging them.[9]

The second culture involved in an exposition is that of the nation where the exposition is held. The organizers have particular agendas that foster certain institutions and ideas at the expense of others. European fairs from their inception stressed the far-flung empires not only of the French and the British, but also of the Germans, Belgians, Italians, Dutch, and lesser powers. Americans at first were far less comfortable with the imperial theme, though they did present aboriginal peoples as part of an evolutionary tale of progress at the Chicago Columbian Exposition of 1894.[10] After the war with Spain in 1898 Americans began to emphasize their overseas possessions more, and they became a common element in fairs thereafter. The San Francisco Fair of 1939 particularly stressed American expansion into the Pacific.[11] In contrast, the New York "World of Tomorrow" made the evolution of democracy its main theme, and commemorated the 150th anniversary of George Washington's presidency. In 1789 New York had been the national capitol, and fair organizers replicated the swearing-in ceremony on the same spot, in downtown New York. In contrast, the theme center of the fair looked ahead to democracy in "The World of Tomorrow." San Francisco emphasized American expansionism, while New York's "Democracity" was a gigantic scale model of the city of the future.

How was it possible for two quite different fairs to be staged simultaneously in one country? The answer lies in how American fairs were organized. Each nation developed its own approach to financing and managing expositions. French fairs were largely controlled by the national government; the British relied more on private financing, but still looked to Parliament for leadership. In the United States the federal government played a relatively minor role. European expositions generally were held in national capitals, but Washington DC has never hosted one. In the United States fair organization has always required a partnership of local leaders and businessmen, who initiate the idea and then seek support from state and national government.[12] Responsibility for running each fair is lodged in a private corporation created for the event. It is charged with raising funds (often through selling bonds), developing the site, recruiting the participants, licensing rights to the world's fair logo, and managing day-to-day operations. The emphasis on local initiative and corporate organization means that businessmen have considerable control, since their financial participation is essential. Put another way, American expositions were not ordained by Congress or the President, who did not offer initial leadership but rather provided legitimacy to an effort already begun. In America it was possible to hold two world's fairs in 1939 with quite different themes, different architectural styles, and different international participants.[13]

The New York fair was conceived in 1935 by three area businessmen, who then got endorsements from leading citizens before taking the idea to the governor and the President.[14] Given this background, it should not be surprising that the New Deal was not greatly in evidence at the New York fair of 1939. Except for one small Works Progress Administration (WPA) building, tucked away in a corner and largely forgotten, the actual national government was less evident than the symbolic federal state, particularly the Washington administration. Because American fairs of the 1930s were dominated by business leaders, they stressed the rationalization of agriculture and industry as the key to material progress. This was to be measured in new experiences and new consumer goods. Each fair predicted the return of prosperity, not through government regulations and programs but through unfettered private enterprise. Rather than join in the widespread search for a usable past which was occuring in the 1930s, these expositions primarily sought a usable future.[15] They contrasted the "empty" America before Europeans had arrived with the rapid conquest of the continent and the promise of a utopian tomorrow.

The New York Fair was the culmination of the whole series of Depression era expositions in Europe and the United States.[16] The last great expression of the interwar world of colonies and empires, it was a tremendous popular success in the United States, drawing no less than 45 million people during its two-year run.[17] Who came? A disproportionate number were professionals and white-collar employees, evenly divided by sex. People between 20 and 44 years of age were over-represented. While only four percent of the population had college degrees, 38 percent of the visitors did. It was a distinctly middle-class audience with the means to travel, and nearly one-half had come to New York from more than 250 miles away. Only one-fourth were from the immediate New York area. No doubt because they often journeyed great distances, more than half "made between two and six visits to the Fair."[18] These multiple visits increased the probability that they would see foreign pavilions.

The fairground (see figure 6) was officially divided into nine zones, but in practice these could be consolidated into four broad emphases: corporations, state governments, foreign nations, and an amusement zone (not shown). As is immediately apparent, national pavilions were on the periphery, while the corporations were closer to the main entrances and clustered around the theme center of the fair, the Trylon and Perisphere. The corporations' central location was one result of a relatively new practice in marketing the fair to potential exhibitors that had been a key to the financial success of the Chicago fair of 1933–34. In expositions of the nineteenth century buildings were devoted to generic kinds of activities and products, with large halls for agriculture, manufacturing, electricity, fine arts, and so forth. In the 1930s, corporations bought the right to create their own buildings, just as foreign nations had done in the past. Even as corporations obtained this privilege, the 1939 fair tended to de-emphasize the individuality of foreign nations, by providing them with uniform exhibit halls. While countries with large budgets could construct free-standing pavilions of distinctive national design, only twenty countries did. The other 38 foreign participants were housed in structures that curved around the Court of Peace. Each flew its flag above the roof-line, and each was provided with a broad shelf about three meters off the ground to hold a characteristic national figure. When considering the expression of national cultures at world's fairs, one must distinguish between the 38 countries confined to the relatively small spaces provided by the architects of the fair, and the 20 who made a greater investment and created their own pavilions.[19] I will focus on the latter group, all of whom came from Europe or the Americas, with the exception of Japan.

These nations did not automatically agree to participate in a major exposition, especially since France held one only two years before. Building a pavilion was expensive, and the world economy was sluggish. Diplomacy at the highest level was employed; President Roosevelt sent out official invitations to the event in 1936. Grover Whalen, head of the

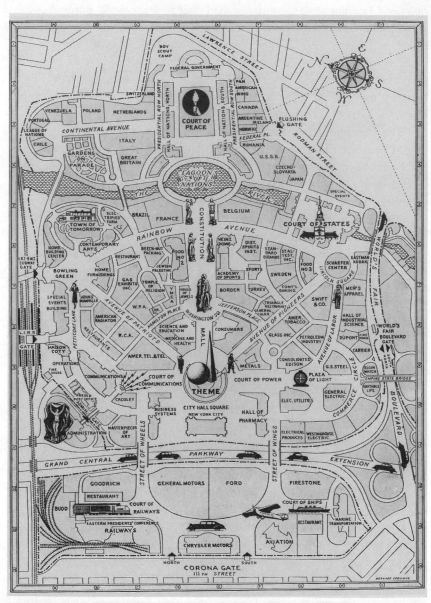

Figure 8: Map of Fairgrounds *(Archives, Smithsonian Institution)*

Figure 9: Court of Peace *(Danish Royal Library)*

World's Fair corporation, then traveled to Europe to drum up interest. In his autobiography he recounts the initial difficulties of attracting foreign exhibitors and how these were overcome once the Soviet Union agreed to participate. "No one was going to allow the Russians to overshadow them," he recalled.[20] All the other great powers except Germany soon agreed to erect pavilions.

Yet after deciding to participate, the projection of a unified national identity is obviously a difficult task, one that requires selection and simplification.[21] Nations are complex, their borders are often arbitrary, and within these borders there are many regions and minorities. What is to be emphasized? Furthermore, the foreign public to be addressed has to be taken into account, for it has its prejudices, assumptions, and ignorance about other countries. Keeping these difficulties in mind, the problem of erecting an exhibit may be divided into three parts: architecture, exhibits, and cultural programs.

II

A national pavilion must be built in a style that can serve a number of contradictory functions. Most obviously, it must harmonize with the themes of the fair, which are usually expressed in an architectural orthodoxy of the moment. In New York, "no imitations either of historic architecture or of permanent materials were permitted."[22] This rule ensured that the overall appearance of the fair would be modern, and it encouraged exhibitors to use either temporary materials such as plywood or gypsum board on light steel frames or new lightweight building materials such as aluminum. At the same time the site offered restrictions of its own. The "World of Tomorrow" was built on a garbage dump that offered a poor foundation. (Indeed, it was built on the "valley of ashes" immortalized in F. Scott Fitzgerald's *The Great Gatsby*.) Consequently, as *Architectural Forum* noted, "the nature of the ground restricted building heights, producing an effect of smaller scale and greater openness than might otherwise have been hoped for."[23] Within these limits the national exhibit had to be distinctive and memorable, suggesting how a particular culture was unlike any other, as the pavilion competed for attention with all the other exhibits.

Obviously, the dangers here are many. A building may blend in so much with the general appearance of the exposition that it is hardly noticed, as was the case with the Irish exhibit, whose unusual Shamrock shape was only visible from the air. Alternately, a structure may stand out too much and look garish, an error committed by the Italian exhibit, with its 200-foot waterfall cascading down the front framed by elongated classical porticos, its gigantic statue of Roma outside, its walls covered with Mussolini's pronouncements, and its central exhibit devoted to a grandiose Italian exposition, planned for 1942 but never held. *Architectural Forum* called the structure, "a curious perversion of classical precedent" that was overly formal and a "heavy setting" for its exhibits.[24] A national pavilion can also err by attempting to blend traditional elements into the fair's dominant style, a mistake grievously obvious in the Portuguese building which looked like a combination of a medieval castle and a small apartment building. Worst of all, however, a national pavilion could over-emphasize well-known icons and become a parody of the clichés about itself. This was the fate of the Canadian exhibit, where uniformed mounties stood guard over totem polls that decorated a prosaic structure.

The balance between architectural distinctiveness and blending in was achieved by several nations at the "World of Tomorrow," notably Poland, Great Britain, France, and Russia. Poland's contribution was not grandiose, but it was architecturally innovative, combining a low white modernist building with a lovely ornamental tower, a semi-transparent lacy metal structure that seemed to be woven together, but which was actually covered with gilded metal plaques. In contrast, the Russian pavilion was a huge, solid structure of marble and granite designed to be shipped back to Moscow and used as a permanent museum. It was dominated by a 259-foot red marble column surmounted by a 79-foot statue of a worker holding aloft a giant red star. As visitors approached the building, they saw a huge bas-relief of Lenin's profile and the words: "The Russian Revolution must in its final result lead to the victory of socialism." The semi-circular building had a wide flight of steps leading up to a central courtyard, which was designed as an amphitheater where moving pictures could be shown. Even without including the "Pavilion of the Arctic," another Soviet exhibit nearby, it was one of the largest and most memorable structures at the fair, a monumental building visible over much of the grounds. In an exit poll during May it rated first among foreign exhibits.[25]

The British pavilion, consisting of two buildings linked by a bridge, was less overtly propagandistic as a structure than the Russian exhibit. Yet, it too was a version of "authoritarian monumentality" which contained a number of large halls.[26] The impressive bulk and sleek horizontal lines of the building gave it a stately yet modern feeling. Outside the gently curving front stood two stylized gilded lions, flanking the doorway, with more lions mounted in bas-relief on the walls. The equally large French pavilion, with a graceful bowed front, cost more than $5 million. Its high portico both gave a glimpse of an elegant interior and served as a restaurant terrace overlooking the Lagoon of Nations. At night it provided a magnificent view of the fair, looking over the waters and the colored fountains. The overall external effect was more inviting and the tone more relaxed than at the British building. Once inside, however, the visitor found that it was something of a maze, with no less than 47 different floor levels, whose complexity made it virtually impossible to produce an intelligible floor plan.[27]

Taken together, these four successful pavilions share certain common characteristics. All were large, easily visible landmarks on the fairgrounds. When viewed from the outside, each had a unity of effect, and did not seem to be subdivided into too many parts. While all were examples of modernism in architecture, none was an extreme example of functionalism or simplicity. Rather, each was complex enough in its external appearance so that precisely what lay within or how the spaces were arranged remained unclear, inviting the visitor to explore.

III

Inside the pavilion an effective exhibit must be created that can perform two quite different tasks. On the one hand, it must create the sense of a bond between the host nation and the exhibitor. This is most commonly accomplished through the display of objects that show a clear historical relationship between the two countries. In 1939 Great Britain provided an elaborate Washington family tree which proved that President George Washington was a direct descendent not only of King John, who granted the Magna Carta, but of nine of the twenty-five barons who had signed it. The founder of the American nation was thus the direct heir of English tradition, and to cement this relationship one of the four originals of the Magna Carta was also on display.[28] A few weeks after the fair opened, the British literalized the metaphor of the tree by planting in the fairground a seedling royal oak from Windsor Great Park, once again encouraging Americans to see their country as a branch of British civilization.[29] William Allen White, dean of American newspaper editors, took the point, declaring in a luncheon speech just before the fair opened, "If I were asked to name the theme of this Fair, I would lead you to the British Building . . .[to] the Magna Carta. It represents democracy and that is the theme."[30] The British were clearly the most successful in asserting a common heritage, partly because they kept the message simple. Holland also stressed its role in shaping colonial America, erecting a statue of Peter Stuyvesant at its pavilion. Venezuela tried to make a more complex international connection by displaying a talismanic object: a lock of Washington's hair that Lafayette had sent to Simon Bolivar. Japan pursued yet another strategy: it made a copy of one of the most important symbolic objects of the United States, the Liberty Bell, replicated in silver and covered with diamonds and pearls.[31] But whether a nation emphasized genealogy, shared values, or respect for the host country's icons, the intention was the same, to forge a link between the two cultures. Fair organizers reinforced this goal by giving each participating nation a "day." June 2, 1939, two and a half years before the attack on Pearl Harbor, was "Japan Day," designated "to stress the cordial relations existing between the United States and Japan," and there were similar days set aside for most of the international participants.[32]

This desire to link national cultures at world's fairs is usually subordinated, however, to the task of projecting national greatness and distinctiveness. There are many ways to do this in displays, but perhaps the most common is to concentrate on a single product. This strategy is particularly favored by smaller or poorer nations who do not have the resources to compete in all fields with powerful rivals, but who can hope to claim a distinctive product as its own province of excellence. The Irish emphasized the "sentiment that lies behind the appeal of Irish linen," tracing the "cultivation of flax to the finished product." There was also a profusion of South African diamonds, Swiss watches, Venezuelan orchids, and Brazilian coffee, each displayed in almost excruciating detail. Venezuela daily flew new orchids to Queens to keep their display fresh. Holland made an even more extravagant gesture, planting one million tulip bulbs in a tremendous variety that harmonized with the color schemes of the different boulevards and theme centers.[33]

Almost as common a way to project national uniqueness is the creation of representative tableaux or spaces that suggest the richness and complexity of another culture. Ideally, visitors can be made to feel, if ever so briefly, that they have entered the other country and imbibed its atmosphere. Often this strategy emphasizes the past, as was the case with Japan, France, and Great Britain at the 1939 fair. Japan particularly stressed its traditional culture and fine art rather than industrial achievements or the war it was then conducting with China, and in this way it sought to appear non-threatening. Its pavilion was modeled on a Shinto shrine, with appropriate landscaping of rock gardens and evergreen trees. Inside was a magnificent grand hall, whose walls were lined with traditional paintings.

The Japanese created a harmony between a traditional interior and exterior. The British building was modern outside but within it also emphasized tradition, in part by serving tea each afternoon. Its popular exhibits included replicas of the crown jewels, an exhibit on the evolution of the Parliament, British heraldry, and, as noted, the Magna Carta. The monarchy was further emphasized by the visit to the pavilion by King George VI and Queen Elizabeth a month after it had opened.[34] At the same time, the British consciously de-emphasized their empire, since its very existence contradicted the major theme of the fair, democracy.[35]

Many nations chose to emphasize their history, though few went to the lengths of the Japanese and the British. The French pavilion, while it contained a few technological elements, was a conscious retreat from the modernism of the Paris fair of 1937. Greenhalgh notes that the organizers, "adjusted the flavor of 'Frenchness' to suit different audiences," and "it was felt the Americans would like something a little traditional. Even though the theme of the show expressly went against historicism, once inside the swishly modern walls of the pavilion (designed by Patout), the visitor was whisked back through five hundred years of French culture."[36] Indeed, they were treated to 11 period rooms in chronological order, with art works by George de la Tour, Largilliere, Rodin, Monet, Corot, Renoir, Degas, Sisley, and Delacroix. The most popular room was a recreation of the young Marie Antoinette's boudoir.[37] Overall, industry was subordinated to fine art, the "scenic beauties of France," and the civilizing force of French culture. While *Architectural Forum* sniffed that the halls were "over-decorated" and the displays "atrocious," the public, less enamored of modernism, flocked to the building.

A quite different and equally effective way to advertise a nation's prowess is through technological displays, which are more effective when they are not static but suggest movement and so begin to embody a narrative. The British only had a few exhibits of this kind, notably Captain Ben Eyston's "Thunderbolt," billed as "the world's fastest automobile," and models of 9,000 ships, representing the merchant marine. The French made more efforts in this direction, including a number of industrial exhibits on the ground floor of its pavilion, such as a Michelin pneumatic-tired railroad car, a diorama of the Le Bourget airport, and several Renault automobiles. By comparison, the Russian pavilion stressed technology almost to the exclusion of all else, and among the 100 displays were a giant catapult machine used to train parachutists and a full-sized reproduction of a Moscow subway station. Through the skillful use of mirrors, it gave visitors a powerful impression of the construction of the lavish underground system, which was far more luxurious than the New York subway many had taken to reach the fairground. Russian exhibits proclaimed the advances made under socialism. "In another hall are dioramas showing the construction of the factory city of Magnitogorsk and of the Ilitch collective farm in the Voronesh region. A system of moving mirrors and changing lights shows life and agriculture in a Russian village today—the disappearance of the church, the transformation of the landlord's estate into a rest home and the coming of electricity and mechanized agriculture."[38] Collectively, the exhibits stressed the strength of Russian science and technology, providing an overall narrative of rapid economic growth.

Just as importantly, this brave new future was presented by attractive young Russian women who, as one commentator noted, were able to "supply welcome clarity to many otherwise confused exhibits."[39] They were, of course, more than interpreters of the exhibits, providing visitors direct contact with people who had grown up after the Russian Revolution. Often foreign exhibitors overlooked the obvious appeal of having representative people from their countries available to discuss the displays and give them immediacy. France understood the importance of such direct contact, and part of the appeal of its exhibit was not just the opportunity to sample perfumes and other expensive consumer products, but to discuss them with chic young women who to many a visitor no doubt represented the acme of sophistication and grace. Similarly, the Russian girl guides were irresistible cultural ambassadors, especially if the visitor was one of the millions of Americans then fascinated with revolutionary politics.

Another common strategy of national representation is to create dioramas or miniaturizations that suggest how national accomplishments are visible in its characteristic landscape. For example, The Netherlands erected a gigantic relief of its countryside, complete with working canals, dikes, and tiny windmills, so that the curious visitor could see the "drama of Dutch land reclamation" as the countryside was alternately flooded and drained. A similar strategy was common to represent imperial possessions. Holland built a sixty-foot-long scale model of the Indonesian countryside, showing representative coffee and tea plantations, as well as rice and tapioca farms. Miniaturizations and dioramas appeared in a majority of the national pavilions.[40]

A related representational strategy is the erection of large globes and maps to demonstrate the wide extent of a nation's empire. The British erected a map of the world that showed the Commonwealth divided into six zones.[41] France erected "France Overseas," a separate building dominated by a giant glass globe showing its network of colonies, mandates, and protectorates. But the most spectacular map of all was in the Soviet pavilion, a 440-square-foot "seven ton map of the Soviet Union, made by lapidaries in Russia for over a year and a half and studded with diamonds, rubies, and many varieties of precious and semi-precious stones found in Soviet territory." Such displays were designed not merely to convey information but to display national greatness and wealth, to over-whelm the visitor with the vast extent of an empire, the richness of its natural resources, and the ingenuity of its artisans.[42]

Another common strategy of cultural self-representation was culinary excellence. France and Italy vied for the honor of having the most elegant restaurant. Turkey and Sweden made distinctive cuisine the focal point of their efforts, by locating their pavilions in the food exhibit area rather than around the Lagoon of Nations. Food literally allows the visitor to ingest part of another culture, often to the accompaniment of appropriate music. The Rumanian restaurant was popular, in part because of its gypsy violin-ists. This strategy can result in self-parody: the Swiss pavilion, along with its huge exhibit of watches and clocks, contained a restaurant featuring chocolate and cheese, consumed to the accompaniment of alpine horns and yodeling.[43]

IV

In contrast to the pavilions and their displays, which were open to all visi-tors for twelve hours or more every day, cultural programs were relatively infrequent, and can be subdivided into types. First, there was the exclusive event for invited guests only, such as a dinner, usually with entertainment. Here the goal was to reward friends and impress dignitaries. Because such an event often forced the host to close the exhibit to the public, it ran counter to the open nature of the fair as an institution. It was used spar-ingly, but was quite common at the inauguration of national exhibits. The French celebrated their opening with a dinner where Lily Pons sang and notable musicians performed.[44] A second, more common form of programming was the public performance of a major work. For example the Hungarians presented Zoltan Kodaly's comic opera, "Hary Janos" in its American debut at the Hall of Music, with singers from the Budapest Royal Opera. Such performances were seldom free, however, and they required that the fairgoer plan well ahead.[45] In this category also were the visiting choruses and marching bands, sent from virtually every participating country and American immigrant group. In contrast to these one-of-a-kind performances, most cultural programs were staged several times a day. By far the most common programming was film, which could be run continually at a low cost. The Coldstream Guards played regularly at the British pavilion for the first two months. The Belgian carillon's hand-cast bells had a beautiful tone that could be heard every few hours.

The pomp of opening ceremonies always received a good deal of newspaper coverage, particularly if important political figures appeared. Mayor La Guardia was usually on hand, along with appropriate ambassadors and representatives of the fair. While people usually gathered to hear speeches and bands, royalty always drew a crowd. Both the Norwegian prince and the Danish monarchs arrived for the opening, but the Norwegians drew the most attention. Their liner rammed and sank the pilot boat that went out to greet them.[46]

V

Any examination of "The World of Tomorrow" immediately reveals that the pavilions of nations were less successful in attracting the public than the exhibits of the great American corporations. This striking fact cannot be attributed to economics, since corporations like governments had diminished resources due to hard times. Furthermore, the history of fairs suggests that for exhibitors, corporate and national alike, immediate profit is seldom expected. Fairs are not about profits but prestige. The anthropologist of world's fairs, Burton Benedict, has argued that the potlatch ceremony of the Northwest Indians is a suggestive analogy to the world's fair, for each requires an enormous outlay of goods in order to impress friends and rivals. The more lavish the expense and the more elaborate the display, the greater the prestige.[47] In the uneasy political climate of the 1930s, European nations had good reason to court public opinion in the United States. Any military strategist could see that should another world war occur, American intervention might well prove as decisive as it had in 1917. Indeed, failure to participate in the potlatch of the world's fair was virtually tantamount to a direct admission of enmity, as can be seen in the refusal of both Franco's Spain and Hitler's Germany to send exhibits to New York.[48] So strong was the anti-Nazi feeling in the United States that briefly there were plans for a cultural pavilion emphasizing German culture before Hitler's rise to power, but this idea was abandoned shortly after it was announced in January 1939. In contrast, the French, the Italians, the Russians, and the British spent large sums and strove to make their exhibits among the most memorable. If realpolitik ultimately dictated whether or not a nation appeared at the fair, once it did decide to participate, few wished to do so half-heartedly.

A better explanation for the greater appeal of corporate exhibits appears to be that they were more cleverly adapted to the taste of the American public. Popular news accounts, histories of the fair, polls by the *New York Times*, and interviews with those who visited the fair, all confirm the over-riding popularity of the exhibits at eight pavilions built by General Motors, General Electric, Ford, Westinghouse, AT & T, Chrysler, the American railroads, and Consolidated Edison. The popularity of these pavilions was not due to their central locations or to their impressive size, but rather to their successful exploitation of theatrical forms of display, their extensive use of miniaturized models of the future, and their reliance on showmanship rather than detailed explanation.[49] Roland Marchand has observed that a clear evolution took place at American world's fairs between the late nineteenth century, when exhibits were primarily educational, and the middle of the twentieth century, when they emphasized entertainment.[50] European exhibits in 1939 appeared old-fashioned by comparison. They tended to be static, demanding that the visitor read and think about them, while American displays were more dynamic and self-explanatory, or they were presented as staged events. General Motors presented the "Casino of Science," and General Electric similarly presented a "House of Magic," where visitors were impressed more than educated. American exhibits often were participatory. Inside the General Electric Hall people could see themselves on the new medium of television. At AT & T one could hear, on a "voice mirror," the sound of one's own "telephone voice," or make a free long-distance call anywhere in the US, with the condition that the crowd listened in on the conversation. The absolutely most popular exhibit was General Motors' "Futurama" (described in the previous chapter) which was besieged by huge crowds who waited for hours to take its journey into the future.

The European pavilions failed to attract as much attention as the corporate exhibits not because they were under-financed or badly sited but because they adhered to older ideas of the representational function of an exhibit. They sought to educate rather than to entertain. They did not simulate an airplane flight over the France or Belgium or Italy of the future, but built conventional exhibits that the public could walk through, or not, as they chose. Europeans put captions on the walls and expected visitors to read them; corporations relied much more on narrators, sound effects, film, and dramatic lighting. European pavilions created static spaces modeled on the traditional museum; corporations synthesized total environments. The full scale model of a Moscow subway station at the popular Russian pavilion was the significant exception. For the most part, Europeans emphasized their fine arts and crafts, while Americans stressed consumer goods, high technology, and interactive displays. Europeans tended to focus on the past, which appealed less to the American public than the corporate emphasis on the future. European exhibits often did not address the central values of the American audience, nor did they employ a mode of presentation calculated to capture American attention. They proved more successful at erecting eye-catching pavilions than in filling them with appropriate displays. If they created a generally positive impression, Europeans nevertheless inadvertently demonstrated that they were not adept at addressing the American mass audience. The corporations, in contrast, showed that they had mastered the forms of popular culture.

The 1939 New York World's Fair showed the importance of cultural diplomacy for both governments and private enterprise. Not only would the neutral United States soon be allied with the countries that had constructed the most popular national exhibits. After World War II, first under the auspices of the Marshall Plan and later within the Common Market, American corporations brought their arsenal of public relations techniques to Europe. The nationalist idea that a culture could be expressed in exhibits about the past would be challenged by international corporations, whose dazzling displays emphasized a new consumerism focused on the future.

NINE
Don't Fly Me to the Moon: The Public and the Apollo Space Program

To many Americans voyage into outer space seemed to represent the continuation of the frontier experience in a new arena. Indeed, as the Second World War came to an end, Vannevar Bush submitted a report to President Roosevelt called *Science—The Endless Frontier* in which he argued that the government needed to fund basic science as an investment in the future. In the 1950s as the "space race" heated up with the Soviets, Americans began to see it as the most important technological competition of the Cold War. President John F. Kennedy's "New Frontier" included exploration of outer space. As he said in Los Angeles when accepting the nomination of the Democratic Party, "From the lands that stretch three thousand miles behind me, the pioneers of old gave up their safety, their comfort, and sometimes their lives to build a new world here in the West. . ." He rejected the notion that "there is no longer an American frontier." On the contrary, "the New Frontier is here, whether we seek it or not", including "the uncharted areas of science and space." Early in his administration he announced a new national goal: to land a man on the moon by the end of the 1960s.[1]

It was an audacious plan. At the time of the announcement the United States had only seriously pursued a space program for about three years. The launch of Sputnik in 1957 hastily called it into being, but the first Vanguard rocket had blown up on the launch pad in December of that year. The Soviets put up the first satellites and sent the first man into space. At the time of Kennedy's announcement they could also launch heavier pay-loads.[2] The United States had a short track record and was behind in the race. Furthermore, even if it were possible to reach the moon so quickly, there was no assurance that the public would back the project. Yet Kennedy had little to lose; he could only stand for re-election one more time, in 1964, long before anyone, Russian or American, could possibly reach the moon.

The Apollo Program was a strange, even paradoxical affair. While there were some supporters whose adulation approached religious fanaticism, it was never that popular with the public as a whole. It is easy to forget this fact when reading the best-selling accounts of individual missions, ghost-written for the astronauts, or some of the earnest scholarly writing trumpeting the scientific knowledge gained. Much of what has been written presents space flight as the most important thing that has happened since life crawled out of the sea onto the land.[3] The most exaggerated claims for the spiritual significance of the Apollo Program were not made by NASA or by the astronauts, however, but by new-age thinkers such as William Irwin Thompson, who argued that experiencing outer space transformed inner consciousness.[4] Yet opinion polls throughout the 1960s showed that the public was hard-headed about the space program, kept an eye on the price tag, and considered other priorities more important. Despite the reluctant public support, however, when astronauts landed on the moon in 1969, an enormous television audience was virtually mesmerized. Apollo XI was a media event.[5]

I

After Sputnik went up in 1957 outer space appeared to be the inevitable "high ground" of the Cold War. Readers of Tom Wolfe's *The Right Stuff* can easily get the impression that the American public was overwhelmingly supportive of the space program. He describes the mass media besieging the astronauts' families, the crowds that greeted them on their return, the ticker-tape parades, and all the media hype.[6] Certainly NASA had a powerful public relations apparatus, and the media lavished attention on the Mercury, Gemini and Apollo programs. A quarter century after Apollo XI some are nostalgic for the heady days of the early launches and call for renewed commitment to exploration. The 1960s depicted in some retrospectives seems to be a time of enthusiastic support, in contrast to the current apathy toward space programs.[7] However, the American public was at best ambivalent about the space project, and became less supportive in the course of the 1960s, well before the astronauts reached the moon.

As early as November 1963, the Grumman Aircraft Engineering Corporation received a study of "Attitudes Toward the Moon Race Among Opinion Leaders and the General Public" which it had commissioned from a public relations firm. Grumman realized that interest in the Apollo program was wavering. As the aerospace firm with the contract to produce the lunar module, it wanted to know if Americans would continue to pay for the program. The report found that support was based more on Cold War fears than the program itself. The "scientific aspect of the program, being largely indefinable in advance of the fact—and perhaps largely unintelligible to the majority even after the fact" had only "some appeal."[8] Thus, even during the Kennedy presidency, when the booming economy promised growth and full employment, about half the population felt that the money "might better be spent on substitutes here on the ground involving general health, education, and welfare." Before President Johnson's Great Society programs, before the urban riots of the later 1960s, and before the costs of Vietnam made the Defense Department an even fiercer rival for NASA funds, already in the fall of 1963 support for the Apollo program was vacillating.[9] Just as important, the public was ill-informed. Less than half the population had "seen or heard anything about the moon race" for "several months."[10] The report concluded that opinion leaders were even less supportive of NASA than the general public.

This preliminary study accurately forecast the growing popular mood against the space program, soon documented by a series of Harris polls. In November 1965 a slim plurality of 45 percent of all Americans favored going to the moon, while 42 percent opposed the idea, with 13 percent undecided. (For comparison with more popular programs, consider that the Civilian Conservation Corps was approved by 82 percent of the population in 1936 and the Marshall Plan in 1948 had support from over 55 percent of both Truman and Dewey supporters, while no more than 22 percent were opposed.[11])

Those most against the space program had a grade school education (61 percent against) and earned less than $5,000 a year (56 percent against). Those most in favor had a college education (59 percent pro) and earned more than $10,000 a year (60 percent pro).[12] Surprisingly, support was weaker in the South than in the East, Middle West, or West, despite the fact that Southern states benefited disproportionately from NASA spending through facilities in Huntsville, Alabama, Houston, Texas, and Florida. Overall support for putting a man on the moon dropped to 34 percent in 1967[13] and remained at that low level during a long period (when it was not stimulated by launches) until February, 1969.[14] The Apollo program commanded support from only one third of the population during the final two years before Neil Armstrong finally stepped out on the lunar surface.[15] The strongest support came from men, the

Table 1: Nationwide Harris Polls, 1965–1969

"It could cost the United States $4 billion a year for the next ten years to finally put a man on the moon and to explore outer space and the other planets. All in all, do you feel the space program is worth spending that amount of money or do you feel it isn't worth it?"

	worth it	not worth it	not sure
Nationwide, 1965	45	42	13
East	45	38	17
Midwest	47	43	10
South	37	49	14
West	50	36	14
By income			
under $5000	28	56	16
$5,000–10,000	52	36	12
$10,000 +	60	28	12
By education			
grade school	24	61	15
high school	47	40	13
college	59	31	10
Nationwide			
July, 1967	34	54	12
February, 1969	34	55	11
July 14, 1969	51	41	8
August 25, 1969	44	47	9

Data from Harris Polls, as reported in *Washington Post*, 1 Nov. 1965; *Washington Post*, 31 July 1967; *Philadelphia Inquirer*, 17 Feb. 1969; *Washington Post*, 14 July, 1969; *Washington Post*, 25 August 1969.

educated, and the wealthy. People under 35 were almost twice as likely to support it as people over 50. The strongest opposition came from African-Americans, women, the least educated, and the poor.[16] The Apollo program had support from only one-fourth of those with only a grade school education.[17] A middle-aged California housewife at a shopping mall, asked if the cost of the moon landing was justified, gave this response: "Well, it gives a lot of people work. It's good for the economy. We should have more money going into the poverty program, but the moon thing will give us power, conquest, world standing and prestige. And I guess that's important because everybody seems to think so."[18] The political point of the "moon thing" was hardly lost on those who were ambivalent about it.

The Apollo Program was most appreciated by those who were young, affluent, well-educated, Caucasian, and male. This seemed to be the group most willing to buy into an ideology that Michael Smith has termed "commodity scientism."[19] The space program seemed justified by the knowledge gained and by the improved commodities "spun-off" as by-products, such as new food concentrates, Teflon, and computer miniaturization. *The Los Angeles Herald–Examiner* made a characteristic list in an editorial: "America's moon program has benefited all mankind. It has brought better color television, water purification at less cost, new paints and plastics, improved weather forecasting, medicine, respirators, walkers for the handicapped, laser surgery, world-wide communications, new transportation systems, earthquake prediction systems and solar power."[20] In making such lists, advocates presented the space program as a cornucopia of practical results of the sort that especially appealed to the well-educated and wealthy.

People in poverty evidently did not believe that things such as solar power or new plastics would benefit them more than direct spending on social programs. The strongest opposition lay within the Black community, where less than one in four people supported the expenditure of $4 billion a year for the Apollo Program. A minority added, "God never intended us to go into space."[21] *The Baltimore Afro-American* had the three Apollo XI astronauts on the front page and called their flight, "Mankind's most thrilling adventure, a spectacular payoff to centuries-old dreams of reaching another planet." But most Black newspapers carried editorials and cartoons attacking the space program, including The *Chicago Daily Defender* and *Muhammad Speaks*.[22] The New York *Amsterdam News* cartoon depicted President Richard Nixon smiling up at the moon while sitting on a huge spherical bomb with a lighted fuse, labelled, "minority frustrations." The accompanying editorial attacked the "outlandish costs of the space race," and declared that, "Man can conquer space, yes. But man has still to conquer his homeland. And that's where the real action is, brother."[23] Some people, black and white, interpreted the event in religious terms. An Indianapolis housewife put her objections this way: "The good Lord never meant us to explore the heavens in the first place." Her use of the word "heaven" instead of space is characteristic of this strand of popular thought.[24] A more typical objection was that of a policeman, who agreed it was "a wonderful human achievement" but who still felt, "it's a lot of money going down the drain."[25]

In July of 1969 on the eve of Apollo XI, the Poor People's Campaign came to Cape Kennedy. To emphasize the slow pace of change, 150 people arrived in wagons pulled by mules. They both protested that the launch was taking place, and, perhaps incongruously, demanded 40 VIP passes to see it close-up.[26] The Reverend Ralph Abernathy urged NASA administrator Dr. Thomas Paine to convert his technology to find new ways to feed the poor. Paine promised to see if food concentrates developed for space could be adapted to feed the undernourished on earth. Paine gave them VIP passes, and declared that the space program was compatible with the war on poverty: "I want you to hitch your wagon to our rocket and tell the people the NASA program is an example of what this country can do."[27] Paine was attempting to harness the old metaphor, "hitch your wagon to a star." But try to visualize what would happen to a wooden wagon hitched to a Saturn V rocket at blast-off. Perhaps African-American views are best encapsulated by a contrast. Duke Ellington, whose music represented an older generation, performed for ABC's national audience a new song, "Moon Maiden," to mark the event. But when the successful moon landing was announced to 50,000 African-Americans at a soul concert in Harlem, many booed.[28]

Modest public support for NASA made it more vulnerable to cutbacks. Starting in 1966, in each of the three years before the moon landing President Johnson sliced its yearly budget, by $370 million in 1967 alone. In part these reductions were possible because the costly initial build-up of facilities at Cape Kennedy had been completed. But there was little public demand for longer-term funding for post-Apollo programs. Just a few days before the Apollo XI launch, syndicated columnist Drew Pearson had these caustic words about the mission: "At Cape Kennedy, the United States is about to launch the most carefully rehearsed, most expensive, most unnecessary project of this century, by which man will reach a piece of drab, radioactive, lava-like real estate hitherto romantic because of distance—the moon."[29] On the day of the launch the Senate Democratic leader Mike Mansfield rejected calls for a program to land a man on Mars, instead emphasizing the needs of people on earth.[30] This proved to be the dominant attitude. Only one month after the moon landing, 47 percent of the population believed that the space program was "not worth it."[31] Strong and even majority opposition thus remained continual throughout the 1960s. Only Cold War fears seemed to justify the expenditures, and these fears were assuaged by a 1967 treaty with the Soviets that demilitarized outer space, by the policy of detente, and (eventually) by the obvious fact that the Soviets had dropped out of the moon race. Despite the spectacular Apollo missions, NASA's budget became a political football.

The public as a whole had a divided reaction to the space program. Most considered it to be a central part of the Cold War, with military implications. About half thought of it primarily as an expensive program that took funds away from social services. Roughly a third valued the program as the creator of many new consumer goods and services, while a smaller group saw the space program primarily as a valuable series of scientific discoveries. In the first years around 45 percent found space travel worth supporting, but by the middle of the 1960s this had dwindled to roughly 33 percent. Enthusiasts often made a pilgrimage to Cape Kennedy to tour the vast facilities or to view a launching. Over the years, those millions who watched spellbound in person were nearly blinded by the light of blast-off, followed by the silent lifting of a rocket taller than the United Nations Building, succeeded seconds later by a mighty roar, as the earth shook beneath their feet. For this segment of the public, an Apollo launch offered a nearly transcendental experience.[32] The majority, however, saw the space program on television, where it had to compete with other programs.

II

During the first years it was often difficult to present a space launch as a dramatic televised event. The failure of the first American Vanguard rocket was a dramatic but embarrassing beginning. The novelty of the early flights drew large television audiences. But lift-offs were often delayed, leaving anchor men desperately looking for someone to interview or some tidbit of information to fill up the time. When the rocket finally went up, viewers were unable to grasp the scale or the impressiveness of the event. On the small black-and-white screen rocket flames were not blinding, the roar was not particularly impressive, and of course the ground did not shake. The flight itself could be followed by a camera, but the sense of speed and movement in relationship to the ground was lost. One had to be at the site to appreciate the scale and power of the experience. On television all lift-offs looked pretty much the same. After the early launches in the Mercury Program, interest faded somewhat. All the networks had was a short dramatic beginning and, much later, a splash

Figure 10: Apollo on its way to the launch pad *(NASA, Washington)*

down. In between, there were shots of mission control, a few interviews, an occasional word on how the flight was progressing, and animated drawings or simulations of what was going on out in space.

This changed dramatically when television cameras and transmitting equipment were installed inside the space craft.[33] There were scientific and operational reasons for including them, but from the public's point of view, suddenly television had a multiple-act drama. After lift-off came a middle segment that featured images of the earth and the stars, weightlessness, space-walks, and interviews with the astronauts inside their ship. With the installation of television in the capsule, space flight became a full-fledged "media event."

As Daniel Dayan and Elihu Katz define this term, a media event is quite distinct from ordinary television viewing. The media event interrupts routine and monopolizes the airwaves on all major channels. Broadcast live and somewhat unpredictable, it is not controlled by the media, although those staging it usually cooperate with television channels out of mutual interest. Usually media events are consciously organized and pre-planned. Examples are the Olympic Games, the Watergate hearings, and ceremonial occasions such as the visit of Anwar Sadat to Jerusalem or the royal wedding of Prince Charles and Lady Diana. (Some media events are unexpected, however, like the fall of the Berlin Wall or O. J. Simpson's flight and police pursuit.) Media events are not numerous, and when they occur they are understood to be historic. Such moments draw enormous audiences and are treated by the media with respect. In the culture as a whole they serve the function of reconciliation. "They are gripping, enthralling. They are characterized by a norm of viewing in which people tell each other that it is mandatory to view, that they must put all else aside." Dayan and Katz go so far as to claim that media events "integrate societies in a collective heartbeat and evoke a renewal of loyalty to the society and its legitimate authority."[34]

The Apollo VIII mission can be used to test this proposition. It was the first to fly around the moon and back, in December 1968. Millions watched as the astronauts sent back images of the moon and of the distant earth. On Christmas Eve, as people were gathered in their homes for their traditional festivities, many watched the broadcast from the moon, and heard Bill Anders read the first four verses of the Book of Genesis. When he wished the distant world a Merry Christmas, most people were moved. A month later the Harris Survey found that Apollo VIII was judged positively by 75 percent of the public. Nevertheless, 55 percent still felt that it was "not worth $4 billion a year to explore the moon."[35] A powerful event,

Apollo VIII won approval and momentarily united people, but this vicarious televised experience did not fundamentally change public opinion. It is probably going too far to say that it renewed loyalty to the society or united all Americans in a "collective heartbeat." Having already paid for the flight, Americans chose to enjoy it, but this did not mean they wanted to keep on paying in the future. Dayan and Katz are correct to identify media events as special in nature, but they over-reach a little in their claims for its social function.

Apollo XI was a larger but essentially similar media event. Certainly it conforms to many aspects of their general definition. It broke through normal programming all over the world, and attracted an estimated 500 million viewers, 130 million in the United States.[36] The majority of these watched in groups, and they treated the moon landing as a festive occasion, not as just another show. While most people did not assemble in crowds at a public arena (although some did witness the moon landing on a large screen in New York's Central Park), homes became extensions of public space, where small clusters of people viewed the event. The desire to watch was very strong, and much ordinary business simply shut down. For example, 8,000 workers at the Western Electric Company near Chicago either left work or never came at all, because they were denied access to television during the lunar landing. On the other hand, Waikiki Beach in Hawaii and the shopping center nearby were as crowded as ever at the moment Neil Armstrong opened the door onto the moon.[37] Yet if some were oblivious to the event, most people watched, regardless of political sentiments or feelings about the Vietnam War. Even at universities that were centers of student rebellion and anti-war feeling, such as Minnesota and Berkeley, the campus seemed deserted, as people crowded into any room with a television.[38]

This interest, which occurred despite an undeniable clash of cultural styles, confirms the idea that a media event welds together the community, momentarily reconciling its divisions. At the launch site old adversaries Lyndon Johnson and Barry Goldwater shook hands. Dale Carter finds that such moments were typical of the "deep-seated, extensive, and voluntary integration" that marked the Apollo XI launch and many of the parades and special events which focused on the astronauts during the 1960s. As Carter rightly emphasizes, from the point of view of NASA administrators such as James Earl Webb, these moments of public integration were intended outcomes, not accidental outpourings of enthusiasm, and they seemed to represent the promise of large-scale technocracy to improve society.[39] Yet to many young people, the crew-cut astronauts with their military backgrounds seemed an obvious part of the "military-industrial-complex."[40] Indeed, NASA administrator, Thomas Paine, called Project Apollo "the triumph of the squares. . . There was no fight from Neil Armstrong when Congress told him to plant an American flag on the surface of the moon. . . The Apollo Program is not only run by squares, but for squares as well."[41] President Nixon loved to be photographed with the astronauts, and he made the most of his opportunity to telephone them at Tranquillity Base from the White House.[42] (Inveterate Nixon haters groaned at that moment, but did not stop watching.) Neither Neil Armstrong nor the other astronauts became student cult heroes. Nevertheless, witnessing the event was a powerful moment of integration, when members of the audience, often despite themselves, were enthralled.

Most still remember precisely where they were and with whom they took that vicarious voyage of discovery to another world. After the launch, after Neil Armstrong clambered down the ladder and spoke his famous line, it was no longer relevant to ask if going to the moon was worth the money. Suddenly that became a rather ignoble question. The moon landing was a paradigmatic example of the media event as conquest, which has two fundamental characteristics: "(1) the reaching beyond known limits through an act of free will, and (2) the resulting charismatic seduction."[43] Dayan and Katz use the term "charismatic" in the Weberian sense, and argue that "Conquests are closest to the rituals of archaic societies" and "represent the eruption of the charismatic model onto a political stage."[44] While this no doubt overstates the case and grants far too much power to a vicarious experience, they are correct to emphasize the distinction between a media event and ordinary programming which is not live and which usually makes no pretense of being historic.[45]

During the early years, when the Apollo Program was perceived as one of many contests within the Cold War, public support often hinged on factors extraneous to the program: costs, party politics, changing national priorities, and competition from other news stories. At the moment of landing itself, however, these concerns fell away like a spent rocket engine. Two men were about to step onto another world, and at that moment all else seemed unimportant. Everywhere friends and neighbors came together and watched the unfolding story. Guards in a county prison became so engrossed in the event that they did not notice the disappearance of 17 prisoners who sawed their way out. NBC estimated that 123 million Americans saw the broadcast of the landing.[46]

The voyage raised tantalizing possibilities. Would the spaceship be too heavy for the lunar surface? Would there be signs of previous life on the moon, such as a fossil or an artefact left by some previous visitor? Could unknown bacteria inhabit the moon and contaminate the earth when the astronauts returned? People jokingly asked if there would be a welcoming committee of little people. Even assuming there was no evidence of life, there was a fascination about seeing the surface of another world. Would it look like a mountainous desert landscape on earth? Would its geology resemble the earth's? Would something unexpected or unimaginable happen? Would the equipment fail and make castaways of the two men venturing onto the surface? Once the audience began to entertain such questions, the landing began to have a numinous quality. People watched in a heightened state of awareness; the astronauts seemed heroic; the moment a break in ordinary time.

Nevertheless, NASA did dampen the excitement by developing a brand of broadcast techno-babble, stuffed with so many acronyms that the dialogue between the astronauts was often unintelligible.[47] Few of the astronauts ever learned to communicate their experience of flight, and none ever did so as well as Charles Lindbergh.[48] Armstrong and Aldrin had few verbal gifts, though Armstrong did produce his memorable "One small step" sound-bite. The most articulate member of their crew, Michael Collins, was left orbiting overhead. The lunar landing was almost unavoidably a media event, but its interest lay primarily in images. For no doubt sensible reasons, the men who landed were chosen as technicians and pilots, not as communicators.

III

Few in 1969 would have believed that in 1997 no one had been to the moon or beyond for a quarter century. To this point I have offered two explanations. First, the public's lack of economic support for the program became a dominant factor after the landing on the moon had been achieved. Given the mood of détente in the 1970s, at the same time that budget deficits began to grow much larger than ever before, it made less and less political sense to spend $4 billion a year to explore outer space. Second, as a media event the first moon landing was not repeatable. After Armstrong and Aldrin's visit, the moon was far less mysterious, and getting there seemed less marvellous each time. Television stations did not plan live coverage of Apollo XIII, and it only became an important news story when disaster threatened. Return trips required new rationales, and after Apollo XVII these were not compelling enough.

There is also a third explanation which in no way conflicts with the other two. Kennedy launched the program as part of the New Frontier, and urged Americans to see the astronauts as latter-day pioneers. But this image remained undeveloped as narrative. There was little sense that anyone was about to take a private rocket and homestead somewhere in outer space, although a few television programs presented this as a possibility in the distant future. But the actual moon remained abstract space; it never became a landscape. Jackson's definition of landscape is worth repeating here: "A composition of man-made or man-modified spaces to serve as infrastructure or background for our collective existence."[49] The public seems to have realized that however much one might admire the feat of sending men there, the barren surface of the moon was inconceivable "as infrastructure or background for our collective existence." Images of collective existence remained earth-bound.

The meaning of the moon landing may ultimately reside in the powerful memories it has engendered. Wyn Wachhorst explains the end of the Apollo Program as a reaction to its very success. To him it was the supreme expression of a civilization, comparable to the pyramids and the medieval cathedrals. At the moment of attaining the goal, on the desolation of the moon, "the only meaningful object was that wisp of color afloat in the black lunar sky, four times larger and eight times brighter than the moon from earth, containing all that we know and are." Arriving on the moon, he argues, transformed our understanding of the earth: "The photograph of Spaceship Earth rising over the dead moon encouraged the shift in collective focus from outer to inner space. . . If the moon landing was the culmination of a half-millennium drive to gain dominion over nature, the view of our fragile world seemed pivotal in turning the quest inward. . . Looking back on Mother Earth enabled some to see her, quite literally, as a living organism capable of death."[50] Indeed, the ecology movement found the image of earth seen from the moon to be a potent icon, which it appropriated from NASA for its own ends. Yet however appealing this new image of the earth, it is surely an exaggeration to say that this visual epiphany turned Americans away from the exploration of outer space. The majority's preference for social programs was already established in the prosperous 1960s. In the more stagnant economy of the following decade, when much of the middle class began to demand curbs on welfare spending, the space program would fare badly.

The astronauts unfurled an American flag on the moon partly in order to claim supremacy on the earth. In retrospect, by 1969 the United States had already reached the zenith of its power and influence, and national self-confidence was beginning to crumble as the nation was wracked by internal turmoils. The Apollo Program itself came to an end in the 1970s, the decade of the defeat in Vietnam, Watergate, the energy crisis, the weakening dollar, high interest rates, and a long period of stagnation in personal income. In retrospect, Americans have found the lunar landings to be one of the most satisfying recollections of the time. The National Air and Space Museum remains the most popular goal of tourists in Washington, allowing anyone to see a replica of the Apollo XI inhabited by the astronauts on their voyage. Just as all Americans revere their Revolution, even though less than half the population actively supported it in 1776, the Apollo Program appears to be gaining sanctity in retrospect. Whatever the future of space exploration, the moon landing is fast becoming a unifying memory.

That memory focuses not on the uninhabitable surface of the moon, which remains mere space. The Apollo Program is recalled with affection because it succeeded as dramatic action, not because Americans could imagine settling this "new frontier" themselves. Yet if Apollo ultimately was understood to be an unrepeatable media event, this was not because technology itself seemed alien to the frontier. Quite the contrary. Patricia Limerick has found "that the American public has genuinely and completely accepted, ratified, and bought the notion that the American frontiering spirit, sometime in the last century, picked itself up and made a definitive relocation—from territorial expansion to technological and commercial expansion."[51]

And of all these new "technological frontiers," none expanded more rapidly or proved more profitable than the computer.

TEN

Postmodernism and the Computer Society

Previous essays have traced how technologies shape the way Americans see the landscapes of Niagara Falls and Grand Canyon, the central role of electrification in the settlement of the West, the creation of an electrified urban landscape, technologies of representation at world's fairs, and the use of energy narratives to explain American development. This chapter turns to the electronic landscape of the computer, or rather how that new interior landscape, cyberspace, came into being. For the computer did not always have a screen full of text and images. Only gradually since the 1950s has it become an important part of how Americans imagine themselves and their society. There appear to be three phases, the first lasting from the end of World War II until the end of the 1970s. Computers were integrated into large institutions, notably banks, airlines, insurance companies, government agencies, and white-collar organizations of all kinds. Also in the first phase manufacturers began to integrate computers into design and production. Second, at the end of the 1970s computers began to emerge into everyday life and consumption, as computer chips were installed in many products. Personal computers began to be sold in large numbers, software was sold ready to use with a minimum of instruction, and machines became "user friendly." The decentralization of the personal computer lasted until the early 1990s, when the rapid spread of the internet marked the start of a third phase. At the moment of writing changes are occuring so quickly that the dimensions and structure of this third transformation are hard to measure.

The introduction of the compuer had social consequences at least as great as those of the railroad in the nineteenth century or the electrical system in the first half of the twentieth century. Indeed, the electrical system was virtually complete by 1940, and quite literally laid the groundwork for the computer. Just as electrification occurred at the same time that modernism emerged in the arts, the computerization of society was simultaneous with the cultural movement known as postmodernism. A precise definition of this term will probably ever elude us, not least because it has been continually redefined by critics since the late 1940s. Ihab Hassan once listed a series of oppositions that suggest the difference between modernism and postmodernism, which he saw as a shift in the arts from design to chance, from centering to dispersal, from metaphor to metonymy, from determinacy to indeterminacy, from purpose to play, and so forth. Historians confronting the late twentieth century cannot avoid noting the simultaneous emergence of the computer and postmodernism, not least because Jean-Francois Lyotard in *The Postmodern Condition: A Report on Knowledge*, and other theorists of the movement do refer to the new information technologies. For example, Lyotard wrote, "What capitalism is doing today is to exploit a force that it had neglected up to now, namely language, thanks to the development not only of the media but also of information technologies. The goal is the computerization of all society, which is to say all exchanges of phrases important for society."[1] Lyotard comes close to equating the emergence of postmodernism with the emergence of the mass media and a computerized culture. Perhaps the computer was not a material precondition necessary for postmodernism to emerge but certainly information technologies did manifest a new relationship between language and the human subject.

The first calculating engines were conceived hundreds of years before the computer, but these were mechanical machines, with moving parts which took up considerable space. Physical constraints limited how many calculations could be performed, in contrast to the microchip. One can trace technical and mathematical developments from Napier, Leibniz, and Pascal to Aiken, Zuse, and Atanasoff. But this is not the place to repeat the pre-history of the computer. Suffice it to say that by the middle 1940s a few large electrical machines that filled whole rooms had been constructed, notably one by IBM at Harvard, which was used by the US Navy during the Second World War. At the war's end computers were ready for commercial development.

In another generation social historians will be trying to understand what life was like before the computer, and what the computer meant for the general population during the great transformation from a society that was self-consciously modernist in the arts and mechanical–electrical in industry to one that was postmodernist and computerized. They will perhaps see the emergence of a large white-collar sector during the electrical period as a precursor of the fully computerized society, when virtually all work became one or another form of information processing. They will note that the computer accelerated information flow and analysis, making possible a compression of space and time. They will perhaps be bemused and derisive over certain theories, fears, and expectations that seemed self-evident between c. 1950 and 1980. For reasons that I will come to later, they will probably revise Michel Foucault's deployment of Bentham's panopticon as a model for the technocratic ideal of total observation and social control. They will note that Jean Baudrillard, among others, complained that the United States was a society characterized by the triumph of effects over causes, and of surface and pure objectivization over the depth of desire.[2] As the following will show, Baudrillard's generalization applies with considerable force to the computer, and not in the ways that Lyotard expected.

Starting in the late 1940s, the computerization of society took two generations. After c. 1980 the computer increasingly seemed to herald a social transformation in which hierarchy would give way to a democratic dispersal of information. Before then, however, the new machines seemed to symbolize centralized control, systematic information gathering, and the invasion of privacy. During the 1950s people increasingly came into contact with computers, both as consumers and in the workplace, and in each area the computer seemed to harden the boundaries of the world as it had been described by such critics of the modernist period as Theodore Adorno and Herbert Marcuse. The computer seemed to promise only the ultimate negation of the individual, making possible an extension of the rationality of the assembly line and its interchangeable parts into new areas. Kurt Vonnegut's 1950s novel *Player Piano* hypothesized a not too distant future where all work would be performed by machines that could be supervised by a few technicians, leaving the vast majority purposeless and powerless.[3] Such views were especially common in the first decades after the Second World War. The computer seemed destined to create new forms of alienation and impersonality. It also lead to many exasperating errors, such as that experienced by a Tennessee farmer, who regularly paid taxes on his property, but discovered too late that a computer in the tax assessor's office had not registered his payments and put his land up for sale.[4]

In the first period computers were used almost exclusively by large corporations and institutions, including factories, banks, airlines, the Internal Revenue Service, and the Defense Department. The military and the CIA funded a considerable amount of the early research, cementing the apparent linkage between computers and top-down social control. The machines of this period became increasingly powerful, more expensive, and more demanding to operate. Using them required extensive training in computer languages, to learn to write programs. A single error in a program thousands of lines long prevented it from running correctly, and computer programmers became known as a fanatically exact group of perfectionists, working in an arcane specialty. Sociologists studying the "machine-man interface" found that a programmer's personality often changed after prolonged exposure to the computer, becoming severely logical and machine–like. During the second, transitional period, which lasted from 1980 until the early 1990s, computers became so much smaller, cheaper, and "user-friendly" that virtually all literate people were able to own and use them. At the same time, the hard-disk memory in these machines doubled and redoubled every few years, and thousands of off-the-shelf applications became available. The secrecy of the first period broke down, as a generation of hackers demonstrated that there were few codes that could not be broken. These two periods can be contrasted statistically. In the United States, between the late 1940s and 1977 only half a million computers were sold. By then, Apple Computer and a number of other companies dedicated to personal computers had been founded. Ten years later thirty-seven million computers were in use, and 50,000 computers were built and sold every week.[5] Moreover, these new machines were becoming far more powerful each year. The earlier computer system consisting of a few giant mainframes, each connected to many terminals, was now overlaid with a decentralized system of personal computers, each more powerful than the original Electronic Numerical Integrator and Calculator or ENIAC machine, and each able to network with millions of other machines.[6]

By the middle 1990s the internet, which originally had linked only a few thousand scientists, had become an international forum for discussion groups, home pages, advertising, self-publication, marketing, collaborative writing, and virtual libraries. The Library of Congress announced a program to put millions of documents on-line, so that its collections would become universally available. Politicians of all stripes agreed that America needed to build an information superhighway, which would democratize knowledge and ensure US leadership into the next century. The computer, once feared as the handmaiden of authoritarianism, was increasingly understood as the guarantor of democracy, the means of universal education, and the key to economic growth. By the middle of the

1990s cyberspace was the hottest topic in the media, subject of a new section in *Time* magazine, and the darling of Wall Street investors.

II

In contrast, between 1950 and 1980 most consumers experienced the computer as an impersonal mediation between themselves and large institutions, symbolized by stacks of computer cards perforated with tiny holes. The cards often came in the mail along with bills, and the consumer stared at them and wondered if perhaps an extra hole in the card could cancel a debt or transfer money to a secret account. The cards were also used at most universities, and students had to present one on the first day of class as proof that they had paid their bills and that they had been admitted to that course. On the back of each card was a warning that one should not bend, fold, spindle, staple or mutilate it, or it could not be run through the machine. In 1969 a graduate student who taught freshman English at the University of Minnesota told her students to tear up their cards rather than hand them to her, which they at first refused to do, fearing the consequences. Only after she assured them that the administration had never asked her to send these cards back to the registration office, did the students rip them up, cheering. Her gesture, their fear, and the emotional release created by tearing them up reflected the general feeling that the computer was part of a vast impersonal bureaucracy in modern life. Characteristically, there were rumors that the federal government had a huge, secret computer, containing information on anti-war protesters and members of left-wing political movements. At the end of this first period, computers seemed the very model of technical domination, the materialization of George Orwell's slogan from the novel 1984: "Big Brother is Watching You." It was no accident that when Lyotard wrote *The Postmodern Condition* in the late 1970s his first pages concerned the effects of the computer on society, and that the implicit image of the computer was that of the first stage. He wrote of the "hegemony of computers" that brought with them "a certain logic, and therefore a set of prescriptions determining which statements are accepted as 'knowledge' statements."[7] Computers, he correctly understood, were a central apparatus in the creation and maintenance of a new circulation of information, but he assumed that this circulation and the machines that made it possible would be exclusively controlled by large institutions. The burden of his book was to contrast this Parsonian and Habermasian world of transparent data, equilibrium, and social adjustment with the narratives and language games of the postmodernists. Lyotard was ineluctably a product of his own historical moment, however, no more able to foresee the swerve in the history of the computer that was occurring, even as he wrote, than were the executives at IBM, who had invested all their energies in producing mainframe machines for big customers.

In the late 1970s, however, the personal computer appeared. At the end of World War Two the United States had one ENIAC computer that weighed thirty tons and filled a large room. By 1977 microcomputers were easily available that were twenty times faster than ENIAC and which used no more power than an electric light bulb. They were about the size of large radios and cost only a few hundred dollars. Because of such miniaturization, the personal computer was possible. The once mysterious machine controlled only by large corporations became available to anyone with a moderate income. By the actual year of Orwellian doom, 1984, attitudes toward the computer were undergoing a rapid change. It no longer seemed so anti-democratic, or so wedded to powerful institutions. Instead, it was rapidly becoming common for every household to have its own personal computer. Not without reason, *Time* magazine named the personal computer as its "Man of the Year." If there was less privacy than before, this condition was universal. In the Irangate crisis, as opposed to Watergate, a great many records could not be destroyed, because they were deposited in a locked memory bank of the White House computer. Even though Oliver North thought he had erased all his electronic files, enough still were in the hard disk to send him to jail. Unlike the Watergate scandal, there was no need for a "deep throat;" the memory banks were quite sufficient.

III

"The parallels between Steve [Jobs, founder of Apple Computer] and Henry Ford were striking. Neither man was educated as an engineer nor invented the technology behind the product that would bring him massive wealth and attention. They were, instead, leaders of a social revolution to empower the common man. Ford envisioned the automobile as a mass-produced tool that would give the average person incredible new freedom to explore the world. Steve saw the same in the personal computer."[8] Thus John Sculley, President of Apple Computer, oversimplified two careers, by pointing only to the product, not the method of production. Ford also pioneered the assembly line, which hardly conferred incredible freedom on the common man or woman, though it did raise productivity, and made

possible both higher wages and large profits from the resulting efficien-
cies.[9] The advent of computers in factories at first seemed to be only an
extension of managerial technical control exercised over workers. From
the 1950s until the 1980s the computer was understood to be a tool which
managers could use to monitor production and human behavior. But it
soon became obvious that computers could do much more than merely
store and transmit information. CAD–CAM, or computer-aided design
and computer-aided manufacturing were developed. The first eventually
made it possible for the designer's representation of a product on a screen
to be manufactured directly from computer-generated instructions.
Computer-aided manufacturing sought to incorporate worker skills and
routines into standard machine operations. The ultimate goal was that
which Vonnegut had ironically described: all worker knowledge would be
translated into text and mathematical formulae that the computer could
use to replicate what workers had formerly done. A few technicians would
replace human beings with robots doing routine tasks and computers
would monitor the behavior of those workers who remained. Starting in
the 1960s, at many factories individual machine tools that once were
controlled by the operator came under computer control, deskilling the
work to be performed. This result hardly surprised, as it seemed to be one
more example of substitution of capital for labor.[10]

Yet as computers became more widespread, they also became easier to
use, and it soon appeared that they could induce other, less expected
results. For the textualization of factory knowledge applied to managers as
well as to workers. By the 1980s Harvard University researcher Shoshana
Zuboff had conducted on-site studies of fully computerized workplaces in
operation, and these began to show that despite management's explicit
intention to use computers to deskill workers and to increase its control
over operatives, something quite different was taking place in some
companies.[11] Computers demystified management, and in the hands of
skilled workers they turned out to be useful in undermining hierarchy,
secrecy, and centralization. A fully computerized plant ceased to be an
opaque system; communication no longer flowed only from hidden
sources at the top down to middle managers, who then allowed only a
trickle of information to reach their subordinates. Computers potentially
gave every operator access to all the information in the system, and clever
workers were quick to appreciate this fact, particularly if it could protect
them in disputes. Previously, both workers and managers withheld a good
deal of their working knowledge from one another, preserving realms of
autonomy. In part they did this intentionally in order to preserve power,

and in part procedures and assumptions were often difficult to verbalize. In contrast, at a fully computerized plant a high percentage of this unspoken working knowledge is transformed into text that provides information about all aspects of the operation that is far more detailed than what was available before.

In theory managers can withhold information in a computerized factory, inhibiting its free flow in the system. But managers themselves found that they were only part of a larger hierarchy of observation and information flow, as each level had to report to the level above. At sites ranging from sawmills to telephone companies they soon discovered that it was difficult to confine the flow of information between different levels of the organization. At this stage, "The adversarial vocabulary of 'us' and 'them' invaded the language of operators toward their managers, and plant managers toward divisional executives. This mistrust was not rooted in a perception of evil or malicious intent. It was, instead, the feeling evoked in the silent dance of the observer and the observed."[12] For managers and workers alike, to have every action recorded automatically in a computer system and available for analysis creates "a sense of vulnerability and powerlessness." Managers began to realize that to develop the full potential of the computer system they had to provide workers with wide access to the system's potential, since the computer permitted the synthesis of hands-on knowledge with detailed management records generated from the computer system. Yet at the same time they "found the new data-rich environment to be a humbling experience" because they saw a new complexity to the production processes they tried to control. As one put it, "If you like to control, then it is frustrating. Data opens up the organization. Everyone has to be more humble, modest, open."[13] In many plants the textualization of work and the resulting transparency of operations and individual performance meant that, "The model is less one of Big Brother than of a workplace in which each member is explicitly empowered as his or her fellow worker's keeper. Instead of a single omniscient overseer, this [computer based] panopticon relies upon shared custodianship of data that reflect mutually enacted behavior. This new collectivism is an important antidote to the unilateral use of panoptic power..."[14] By the end of the 1980s it had become clear that the "smart machine" which for thirty years had been expected to concentrate power and information, instead would disperse both through the company. It would weaken the position of managers and empower workers. Nor was this merely a theorist's vision. Zuboff's *In the Age of the Smart Machine: The Future of Power and Work* appeared to rave reviews in the business community, and it was selected by

Business Week as one of the ten best books of 1988. *The New York Times Book Review* gave it front-page treatment, calling it "a work of rare originality." By 1990 Zuboff was on the lecture circuit abroad, sponsored by *The Economist.*

One might object to Zuboff's argument: computers may not always be used to break down manager–worker divisions, since they can still be used to standardize and deskill many tasks and to give management a late twentieth-century panopticon of power. Indeed, in 1990 both approaches were common. The computer, like any other machine, is not autonomous. It opens up certain social and economic possibilities, but which possibilities are selected varies. Indeed, in some cases, the choice is abstinence. Just as the Amish in Pennsylvania reject electrification, some writers refuse to use word processors. People make selective use of a technology. Yet a corporation must make its choice not in the abstract but in a competitive business environment, where only the most efficient survive. Given the increasingly segmented markets of the 1990s in which demand is shifting away from mass-produced goods to smaller batch production, computer-literate workers prove more efficient than those performing rigid routines. The textualized, transparent factory emerges as a very competitive option, reinforcing a trend that is already underway toward more teamwork and democratization in the workplace.[15]

Yet this new collective sensibility is founded upon the interiorization of management's values, which are embedded in the programming of the computer itself. The capitalist imperative to be efficient dictates that the new transparency be exploited to the limit, and companies which refuse to do so eventually cease to be competitive. In undermining management's once exclusive access to information, therefore, one must not imagine that a fully computerized plant undermines management's values. If it empowers those who once had a narrow view of their job and did not understand the interconnections between their work and that of others, this result is achieved as workers embrace the managerial values of maximum productivity, efficiency, and elimination of waste.

IV

The capitalist, computerized factory thus implies a new psychic economy in which secrecy is replaced by openness, privacy by exposure, individuality by teamwork. Emotionally, this is a cool world, for in data-rich work environments supervision no longer requires much face-to-face contact. Some foremen "supervise" people they have never seen but whose work they have directed and evaluated through their terminals. Corporations discovered that often the worker no longer needed to come to the office, as tasks could be sent to a home terminal. Yet if immediate personal contacts are increasingly mediated, it is impossible to hide from the pervading system. For example, it is possible in large operations to monitor secretaries working at word processors, noting the number of keystrokes per minute and the number and length of their coffee breaks. A sophisticated system can also account for the difficulty of the work, and what had previously been subjective managerial judgements about performance become a mathematical "fact" available to all. Computerization thus "has not only provided workers with the language to confront their managers but also equalized their respective realities, since the objective record stands as final arbiter of what has happened."[16]

The computer creates an unceasing flow of information, which becomes the only validated reality. The perception of a fellow worker as being efficient or lazy no longer matters, as the individual becomes the sum of his textualized achievements. In translating the total operation of the factory into an ongoing discourse, the computer encodes a complete simulacra of the factory world. It becomes a decentering machine, de-realizing the shop floor and the boardroom and replacing them with a vast, transparent text. And just as the postmodern world is characterized by a speeding up of image production, saturating the citizen with all kinds of information, the discourse of the new factory has been designed with the goal of speeding up communication, which becomes the analogue of speeding up production. People become accustomed to ever faster psychic turnover, not only in the blizzard of imagery on television but in the pace of corporate discourse, and managers report that where they once could sleep at night they now feel obliged to call in, for the discourse is being created continually around the clock.[17]

Such pressures were rare in the former Soviet Union and Eastern bloc countries before 1989, despite the attempts of some critics to conflate capitalism in the West with what they term the state capitalism of the East. As recently as 1985 Michael Shallis could write in a book on the computer revolution, that,

Technology [primarily the computer] springs from a particular attitude toward the world that is linked closely to the developed countries' form of economy, namely capitalism, both competitive, market capitalism epitomized by Japan and the USA, and state capitalism as practised in the Iron Curtain countries. Such economies are based on a technology that separates information from skill, and which always seeks either a well-controlled work-force or its replacement by more reliable machines. Thus the new technology enables technocratic control to be exercised ever more strongly and offers another means of the replacement of the 'human factor' in the wealth generation process.[18]

Such arguments clearly were dated five years later. The disarray of Russia in 1990, where the rouble had virtually ceased to be the real currency in many transactions, where most goods were not reliably available in the stores, and where industrial plants were hopelessly outmoded, made an oxymoron of the term "state capitalism" which evidently cannot foster an advanced economy on a level with Japan or the United States. Shallis's argument would have seemed an appropriate critique of the American production system in 1950 or 1960, but it is simply irrelevant in the data-rich work environments of the 1990s. Highly computerized corporations simply do not attempt to separate information from skill, nor do they find it profitable to control workers in the same ways as before. Such practices and the kind of alienation they involved, belong to a previous epoch.

Future historians will perhaps return to the list of oppositions compiled by Hassan in 1985, to see to what extent the computer participated in the shifts from the humorless purposefulness of modernism to postmodern play, from finished works to performance art, from centering to dispersal, from genre to text, from the master code to the idiolect.[19] In the textualized workplace the common experience of work increasingly becomes that of participating in an endless flow of information, where boundaries of authority become diffuse, and where hierarchy gradually erodes, to be replaced by divisions of work based on objectified skill, as recorded in the discourse generated through the computer. In such a world distance from the final product may be experienced not as alienation but as a contemplative necessity of the system, which emphasizes an endless overview of unfolding discourse, and provides the kind of Olympian perspective once reserved for managers alone.

Yet this new viewpoint will hardly be that summit of knowledge sometimes referred to as "the Enlightenment project." Rather, for those who accede to the new order, it will provide the pleasure of contemplating a vast operation that is visible and comprehensible in every detail, rather like watching a busy airport from a control tower. Occasionally, however, a plane crashes. There will be times when the actual universe is so out of line with its representation, that it impinges on the textualized world of the workplace, in the form of floods, storms, strikes, epidemics, mass enthusiasms, and sudden economic depression. As much as possible, these outside contingencies will be textualized as well, so that if they cannot be avoided, they can be planned for and inscribed within the discourse of production. The ideal of the system will be to achieve a perfect fit between experience and the continuously generated text of the workplace. This discourse in turn will nest snugly within the larger discourses of the evolving market-place, national economies, and international trade. The "fit" between experience and discourse will not be judged by outsiders, of course, but will be generated in a system of continuous feedback and adjustment, as the workforce participates in both composing and reading it. Most external controls, such as those infamously instituted by Henry Ford in the later 1920s and 1930s, will no longer be necessary. The textual system will continually introject into the workforce the values embedded in the meta-text of production, and they will come to judge their experience and achievements by their congruence with it. Thus, the transformation of the workplace into a continuously developing text signals the emergence of a new kind of psychology, for the computerized workforce will be consti-tuted through an interactive system rather than a hierarchical one. It will interiorize the values of management, which will be embodied in the logic of the programming. This shift in psychology is already expressed in a new vocabulary and new metaphors for a decentered self that can be continu-ally reprogrammed, and which inhabits an indeterminate universe. Diane Ackerman noted in 1990: "For a while neurologists railed against comparing the brain to a computer, because it seemed terrifyingly auto-matic, amoral, and mechanistic. Now the computer simile is back in vogue, because the similarities are so obvious as to be undeniable. The brain is the computer; religion, prejudice, bias, and so forth are all soft-ware."[20] And software does not represent any fixed truth, of course, since it must be constantly updated.

While the new work discourse gives the pleasure of the comprehen-sive gaze, it also creates psychological problems. Most obviously, all personal routines can become subject to continual exposure and review, and the cultivation of habit can only be permitted temporarily, contingent upon whatever new rationalities and refinements may be introduced into the system. If privacy, subjectivity, and desire are repressed in the realm of work, they will find expression elsewhere. Likewise, the easing of hier-archy and the substitution of numerical controls for personal authority will not signal the disappearance of charisma and leadership but displace them to other realms of life. What is erased from the workplace must be inscribed in the heterogeneous and discontinuous text of consumption, in the leisure of multi-channel broadcasting, in the autonomous empire of signs deployed by advertising, or in Umberto Eco's hyperrealities. Computerization thus rewrites the spheres of production and consumption as two distinct discourses, the one devoted to absolute transparency, the other necessary to express repressed or excluded emotions in the realm of popular culture and leisure.

There remain two difficulties in the logic of this textualized system: (1) criminal abuse of the discourse of production, and (2) the discrepancy between the transparency and purposefulness of work, and the play of incongruities in the world of leisure. The model Zuboff proposes takes little account of criminality. While one must be wary of placing too much emphasis on scattered examples, consider the following three cases, all from near the end of phase two, during the fall of 1988.[21] In October, Donald Burleson, a programmer, planted a virus in an insurance company's computer as revenge for being sacked. The virus wiped out 168,000 records, illustrating how dangerous firing people can be for a computerized firm. In November a Cornell University graduate student released a virus into a nationwide computer network, affecting 6,000 machines, including some owned by the Defense Department. His father had been part of the "Core Wars" computer game group at AT & T Bell Labs in the 1960s, in which programming experts had attempted to design viruses and hacker strategies. The son had released a virus that automatically filled up all available memory space, so that machines choked on meaningless data. In December, Kevin D. Mitnick, 25, was held without bail after he gained illegal access to Digital Equipment Corporation computers, and performed a series of tricks that cost the company $4 million. Mitnick was denied unsupervised access to any telephone while in prison, for fear he might re-enter the computer system. These three cases, all from a period of but three months, suggest that the discourse of the workplace will not remain untainted by eruptions of anger, mischeviousness, and criminality. The first case suggests that when a computerized corporation makes employees expendable it is potentially exposed to attack. The second case suggests how easily a form of play encouraged by computerized systems can become dangerous. The third case emphasizes how vulnerable computer systems are to skilled outsiders, even when they are held in prison. Aside from errors that inevitably must creep in, computerized discourse can be erased, infected, or tampered with, and to the extent that the world is linked to this discourse, the world will also be changed. Thus a bank teller may transfer money from customer savings into a secret personal account and then bet large sums on the horses, not only defrauding his customers but changing the odds for bettors at the race track. More recently, Nick Leeson, a young investment banker working for a British bank in Singapore, began to invest huge sums in the Japanese stock market. Because of inadequate supervision, he was free to lose so much that the bank collapsed and went into receivership. Others may do serious damage with less intent. A hacker who taps into a military computer may think he is only playing, but the army will go on alert.[22] Transforming work into discourse may well dissolve managerial authority, in short, but it also opens new arenas for manipulation and transgression, and it permits a continual rewriting of experience.

What these individuals have done is import into work a playfulness that is only sanctioned in the realm of leisure. The teenager who almost starts World War Three by tapping into the Defense Department's computer system does not realize that his "game" of thermonuclear war is "for real." The father plays at "core wars" as part of his job, and his son, doing the same thing, infects 6,000 computers. The new psychology of computerized work may prove to be incongruent with the psychic economy of computerized leisure, and these discontinuities will necessarily erupt into the textualized world of production. As David Harvey has noted, in the arts such incommensurabilities lead to techniques such as collage that juxtapose contradictory realms, or to works that explore alternative social worlds which often collide with destructive consequences.[23] His example was the film, *Blue Velvet*, whose main character moves between two completely incongruous realities in the 1950s, but the same may be said of many other recent films, such as *Diva*, in which a young opera lover makes an illegal recording of his favorite singer at a concert and accidentally receives another tape valuable to the underworld, and his life is spooled into another realm. Significantly, most such films explore contradictions outside the realm of production. They seldom depict the central contradiction between the textualized visibility of a computerized workplace and the glossalia of the consumer society.

The postmodernist debate, in which Lyotard, Foucault, and other Continental thinkers played such a large part, and in which American academics took great interest during the 1980s, failed to describe its own immediate future. Rather, it was part of the transitional phase between modernism and the next social stage, in which computers will play a role that neither they nor corporate executives could envision in 1980. Ever since Henry David Thoreau ridiculed the construction of a telegraph from Maine to Texas, machines have been attacked as improved means to unimproved ends. But some new technologies make possible new ends. Just as the process of electrification created a vibrant new commercial landscape, the process of computerization has democratized information in ways that neither workers nor managers anticipated. The rapid emergence of the Internet in the 1990s has given Americans new ways to decentralize and democratize marketing, publishing, research, access to information, and public debate. Just as important, it advances the textualization of society.

Postmodernism emerges, then, as a phenomenological description of late twentieth-century art and leisure during the creation of an electronic society. Postmodernism is the inverted face of a production and consumption system in transition; the hallmark of that transition is the creation of an entirely new realm: cyberspace.

CONCLUSION

Technology and the Construction of American Space

How have Americans used narratives to understand the role of technology in both the literal and the figurative construction of American space? It would be disingenuous to answer this question with a straightforward realist narrative. Instead, keeping in view questions of narrative structure, I have presented ten case studies of technology in American space. All ten are primarily or exclusively concerned with the twentieth century. Taken collectively, they provide examples of at least six characteristic narratives. In the first of these, machines may be largely overlooked, as in the Turner thesis. In such arguments, when technologies are mentioned, they are taken for granted as being "natural" outgrowths of society. Second, machines also may be presented as agents of social amelioration, as in many world's fair exhibits or in the New Deal's vision of rural electrification. Third is the narrative of technology as a means of social control, which is often buttressed with the work of Michel Foucault. Fourth, new forms of transportation, image-making, and communication may be presented as reshaping the perception of space and time, as in media events or in the contemporary perception of the Grand Canyon. The fifth narrative is satiric: technologies lead to unexpected outcomes. For example, the computer is brought into a factory in order to impose more control on workers but in actual use tends to undermine corporate hierarchies and to democratize access to information. Finally, a sixth, apocalyptic narrative depicts new technologies (e.g. nuclear power) as agents of doom. All of these narratives can be used to present technologies as deterministic forces, which, depending on one's assumptions, can lead to automatic growth, social betterment, massive surveillance, transformation of the lifeworld, ironic reversals of intended results, or apocalyptic destruction.

All of these narratives take for granted a conception of space that was fully developed by the time of the Enlightenment. Expressed as perspective views in paintings, as survey lines on the land, and as topographical maps, this sense of space as empty, neutral, and available for use seemed particularly apt as a way to impose order on the New World. The misleading sense of determinism so common to technological narratives is fostered in part by the Cartesian sense of space, which seems to predicate linear developments seen in long perspective. In this neutral space, the world is raw material, waiting to be worked up, and it offers only temporary resistance to what seems an inexorable sequence of events. The narratives of technology all too easily become the counterparts to what Albert Boime has called "the magisterial gaze," a "diagonal line of sight...taking us rapidly from an elevated geographical zone to another below and from one temporal zone to another, locating progress synchronically in time and space."[1] This Cartesian sense of space excluded Native Americans, not only because they seemed to have left little trace on the land, but also because the logic of perspective vision allowed only one privileged point of view, implicitly that of the European gazing west. To the extent that Native Americans were recognized, they were another part of Nature, waiting for development.

In practice all of the narratives considered here often have been deployed as stories of autonomous technology. Yet determinist narratives are inadequate as explanations and dangerously misleading. Human beings, not machines, are the agents of change, as men and women introduce new systems of machines that alter their lifeworld. They quickly come to see new technologies, such as electric lights, space shuttles, computers, or satellites, as "natural." At the same time, the lifeworlds constructed with older objects (as in *The Home Place*) begin to slide toward incomprehensibility, as those who created that landscape pass away. Machines are not "things-in-themselves," but express culture. The meaning of an individual machine can be found both in its specific location and use and through characteristic stories about it. Theories of narration can be used to understand such major events as electrification, the adoption of the automobile, the 1970s energy crisis, or the Apollo Space Program.

The inseparability of machines and modern narratives further suggests that most fiction has a technological underpinning, even if it is not explicit. As Machery put it, "We should question the work as to what it does not and cannot say, in those silences for which it has been made."[2] This idea underlay the argument (in chapter 5) that among other things *The Great*

Gatsby expresses the contradictions between two systems of energy. Competing technical systems introduce discordant figures of speech, metaphors, and ideas, creating contradictions that texts strive to overcome. This may be made explicit, as in John Steinbeck's *The Grapes of Wrath*, which is underpinned by competing conceptions of agriculture: the family farm vs. agribusiness. More often, as in *Gatsby*, technological contradictions are implicit.

By the middle of the nineteenth century mechanical systems had become the central subject for some narratives and a part of the ideological underpinning of many others. And as discussed in the introduction, technologies were incorporated into the major story, against which the others are defined, which constructs America as empty space available for transformation. The familiar version of this narrative is the arrival of early European immigrants to found Jamestown, New Amsterdam, or Plymouth, followed by their gradual penetration into the continent. This narrative, which long began high school history survey books, was also a central part of the colonists' own understanding of their experience. William Bradford wrote of his fellow Pilgrims that they arrived in "a hideous and desolate wilderness, full of wild beasts and wild men," where, he emphasized, no convenient technologies awaited them: there was no house or inn to open its doors to them as they disembarked for a severe winter.[3]

The continent was not virgin land, of course, but an area where Native Americans deployed their technologies of hunting, trapping, farming, and building, some of which colonists adopted. For Native Americans, space was not Euclidean abstraction, it was alive. European Americans saw the world in instrumental terms, and they transformed the space of the New World through the successive introduction of the sailing ship, the plow, domestic animals, the water mill, axes, and firearms. As Americans recreated the landscape, they invented stories about settling the region. In these tales many objects were taken for granted as "natural" elements, such as the saddle, rifle, and metal tools. Yet these complex technologies were each the product of centuries of development. In many colonization stories the machines essential to settlement were either decontextualized (the rifle without its production) or ignored altogether. The result of this decentering was a myth of national origin that emphasized the confrontation between man and "nature," rather than a conflict between races with different technologies. This is the tale that Turner told so well. The apparently logical unfolding of the settlement of the West intertwined technology with Manifest Destiny and the spread of democracy. Underlying Turner's argument lay the (mis)conception of America as largely empty space. This is also the idea that Nick Carraway evokes at the end of *The Great Gatsby*, as he imagines Dutch sailors looking at the supposedly untouched "breast of the New World."

While some Americans embraced the myth Turner proposed, in practice many knew better. Settlers immediately understood that they needed transportation, tools, water, light, and power.[4] The settlement of the West was a conquest in which machines were both prominent and necessary. In this second narrative about technology in American space, progress follows the paths opened up by the railroad, the irrigation ditch, the electrical line, the highway, and the computer. This idea was strikingly visualized in an 1868 Currier and Ives print, "Across the Continent: Westward the Course of Empire Takes its Way." It depicted a railroad rushing across the plains toward the mountains of the West, with agriculture and small towns springing up in its wake. As Michael L. Smith has noted, this image provided "a grid on which Victorian-era Americans could locate the emblems and the direction of their nation's progress."[5]

A related form of this narrative was celebrated by the organizers of the 1904 St. Louis Exposition, as they marked the centenary of the Louisiana Purchase. They emphasized mining and metallurgy, which had been crucial to the region's development, in an exhibit hall large enough to house four football fields. "The great aim of the exposition authorities was to show. . . not alone products and results, but the processes and stages through which the products pass in order that they may become a benefit to mankind."[6] The exhibits included working models of coal mines, gold mines, iron mines, turquoise mines, copper-refining, and the Bethlehem Steel rolling mills. There was also a detailed presentation of the history of brick-making. Over the whole scene towered a 60-foot iron statue of Vulcan. Nor was this all. Outside the Mining and Metallurgy building extended a 13-acre "Mining Gulch" that contained full-scale exhibits: a cement plant, oil well, gold mill, and copper smelter, all in operation.[7] In this story the settlement of American space is inseparable from the technologies employed, and as its logical extrapolation, every twentieth-century fair projected a high-technology future. The successful promotion and naturalization of the electric landscape at a succession of fairs after 1881 exemplifies this process. Ecstatic predictions of technology's coming benefits were central themes of Chicago's "Century of Progress" in 1933, New York's "World of Tomorrow" in 1939, and the 1964 New York World's Fair.[8]

Exposition displays often became the basis for new urban design. Fairs before World War I promoted a neo-classical revival still visible in many public buildings, and more recent expositions inspired such sites as Disneyland, Stanford Industrial Park, and the retirement community of Sun City, Arizona.[9] As these examples suggest, Americans were long willing to wipe almost any site "clean" and start over. Wilderness areas and national parks are only an apparent exception to this generalization. Parks are the points where the narrative of national development symbolically begins. They are valued as the last examples of virgin land, the original clean slate. By returning to them Americans are seeing where they began, in a secular version of the myth of the eternal return to an initial moment of perfection.[10] By holding parks and wilderness areas inviolate, the rest of the landscape is, by implication, ready to be redeveloped.

Grand Canyon could be preserved, in other words, but government planners could also declare in 1946 that:

> Tomorrow the Colorado River will be utilized to the very last drop. Its water will convert thousands of additional acres of sagebrush desert to flourishing farms and beautiful homes for servicemen, industrial workers, and native farmers who seek to build permanently in the West. Its terrifying energy will be harnessed completely, to do an even bigger job in building the bulwarks for peace. Here is a job so great in its possibilities that only a nation of free people have the vision to know that it can be done and that it must be done. The Colorado River is their heritage.[11]

In this vision, the river's meaning is reduced to use value. Its water exists primarily for man, and its energy is to be completely subjugated. The end of World War II is invoked in the image of beautiful farms and homes for servicemen, and the development of the basin becomes part of creating "bulwarks for peace." But the passage goes much further, suggesting that only a free democratic people could sustain such a vision and grasp it as a historical necessity. The Colorado River functions not only as a wild energy source to be subdued and rationally developed. The complete utilization of the Colorado is a part of America's historical fate, an actualization of its Manifest Destiny. A free people will see that they must dominate nature. To realize their freedom, they must recognize, paradoxically, that they have no choice.

The long-term actualization of the second narrative (of human expansion through technology) can also be illustrated by the de-concentration of American cities, discussed in chapter 4. As electricity spread beyond the urban fringe, population density declined. From 1920 on suburbia continued to grow, and after c. 1965 rural zones also began to be repopulated.

Commerce has followed and fostered these trends, as malls, office towers, and corporations move into the hinterland. Americans seem intent on abandoning the European form of the city, even if rearguard gentrification reclaims some historic neighborhoods. An interesting variant on this pattern emerged in the West, where electricity was in place before a large white population had arrived. There it became particularly clear that Americans used technologies to escape, to open new spaces, to create new landscapes, to shape new starting points.

This expansionist story is forcibly contradicted by narratives of social control in which advanced technologies and the pscyhology they foster lead to centralization of power. Jacques Ellul presented a general book along these lines that was widely read in the 1960s.[12] Donald Worster's *Rivers of Empire* provides a somewhat different example of such an argument (discussed in chapter 2), in which the need to control water inexorably leads to the social dominance of an elite. Another widespread version of this narrative focuses on communication technologies. First made popular by George Orwell in his post-war novel *1984*, the sense that "Big Brother is Watching You" seemed increasingly plausible as electronic miniaturization made eavesdropping on telephone conversations easier, and as computerization spread through the society, allowing other forms of monitoring. This narrative has recently gained not only plausibility but a pedigree from the work of Michel Foucault. He presents the invention of the prison, the hospital, and the mental ward, as exemplary studies in the construction of modernity.[13] Following Foucault, one can argue that the panopticon is the model of modern surveillance. His powerful descriptions of social control are often cited in sociological and architectural studies of cities. In a parallel development Mike Davis's *City of Quartz* presents recent public architecture in Los Angeles as a system of fortresses and barriers designed for exclusion and social control. He concludes that in Los Angeles, "genuinely public space is all but extinct" replaced by an "archipelago of Westside pleasure domes—a continuum of tony malls, arts centers and gourmet strips" which is "reciprocally dependent upon the social imprisonment of the third-world service proletariet who live in increasingly repressive ghettoes and barrios."[14] Davis does not imagine totalizing, all-powerful systems to the same degree as Foucault, who tends to slight the creative resistance of residents, workers, and inmates. Actual prisons and hospitals never achieve the totalizing domination Foucault describes. Likewise, the computerized workplace, as Zuboff's case studies showed, cannot maximize efficiency using the panopticon for inspiration, and factories must permit considerable individual initiative and self-control.

Most of Foucault's *oeuvre* is filled with a claustrophobic sense of confinement within limits and surveillance from above that is often ill-adapted to understanding the development of specific American environments. Spaces of consumption, such as world's fairs and shopping malls, are of another order altogether, even if they do have security systems.

In contrast to the scenario of technology inexorably leading to centralized structures of control, in the fourth narrative communication and transportation technologies compress the *experience* of space. Journeys that once required weeks and that necessarily brought the traveler into touch with local residents all along the way, have been replaced by much faster, more comfortable, but anonymous trips. This space–time compression has the effect of transforming what are distinctive places for local residents into increasingly abstract space for those who pass through. Wright Morris was acutely aware of this effect when he drove through the South with a camera in the 1930s, and for that reason distrusted the images he made there, preferring to return to and record his home place.

Yet few observers are as sensitive or as meditative as Morris. Increasingly, the American relationship to landscape has shifted away from the contemplative view of someone immersed in a scene toward the interactive, high-speed encounter. This shift was not only registered in the changes in tourism to national parks, as described in the first chapter on Niagara Falls and the Grand Canyon. The newer sensibility is also manifest in a wide range of sports that became popular in the twentieth century: skin-diving, water-skiing, white-water rafting, parachuting, car racing, down-hill skiing, and many more. As J. B. Jackson has noted, each of these provides a sense of speed, a "sense of danger or at least of uncertainty, producing a heightened alertness to surrounding conditions." None of these new activities allowed "much leisure for observing the more familiar features of the surroundings." These sportsmen have a new relationship to landscape quite unlike the slower-moving cyclist, hiker, or amateur naturalist, all of whom were common figures a century ago. The new landscape is more abstract and "seen at a rapid, sometimes even a terrifying pace."[15] Since Jackson made these observations, new activities have emerged that confirm the trend, such as hand-gliding, snow-mobiling, and roller-blading. People demand speed and immediacy, a maximum of experience in a minimum of time. Vision becomes central and the nuances of feeling through the other senses are often ignored.

The automobile was clearly a part of this general shift in awareness. The railroad had already accustomed people to passing through regions without being able to smell the plants or hear the sounds of the countryside, but the automobile gave each driver many choices, making it possible constantly to shift destinations, routes, and speeds. As with the new sports that emerged in the twentieth century, driving focused perception on an activity in the landscape, rather than the landscape itself. Furthermore, roads increasingly ignored the irregularities of local topography, as highway engineers blasted through hills, bridged valleys, and otherwise reshaped the land to conform to their specifications. The compression of space, with its emphasis on speed and vision, encourages Americans to see landscape as generic space, which it is possible endlessly to rewrite.

Communications technology can easily be inscribed within this narrative of compression. Ever since the introduction of the telegraph, sending messages at the speed of light has been said to annihilate space and time.[16] In the 1960s Marshall McLuhan popularized the notion of a "global village" based on television.[17] In the middle 1980s Joshua Meyrowitz published a well-received book asserting that electronic media have inexorably undermined our sense of place and collapsed distinctions between inside and outside, public and private, civic and corporate.[18] In part, he saw this as a positive development that weakened authority, fostered women's liberation, and raised Black consciousness. Both McLuhan and Meyrowitz presented media as a complex, all-encompassing technology.

The current version of this narrative focuses on the internet. What appears on the computer screen seems a curious combination of space and story, mind and matter, fused together in a location that exists nowhere and everywhere. Cyberspace seems a virtual virgin land produced by technology. Appropriately, all the national parks now have home pages, including portfolios of digital images. Virtual tours of other famous sites have also become available on the internet, including the Hoover Dam and New York City. Those who celebrate the internet generally present it as an inevitable compression of space and time. Nicholas Negroponte writes, for example, that "Like a force of nature, the digital age cannot be denied or stopped."[19] Similarly, Newt Gingrich enthused over the beneficent effects of computers in his *To Renew America*, which relied heavily on Alvin and Heidi Toffler's idea of a "third wave" of technology.[20] In this narrative a revolution in consciousness and society is brought about by a technological determinism which paradoxically leads to more human freedom.

Compression is aided by substitution. The electrified city had been the first hint of the possibility of two or more landscapes occupying a single space, and the dramatic lighting of world's fairs and natural sites were early examples of the merger of nature and culture into a synthetic realm. Just as artificial light once mingled with America's most cherished natural landscapes and man-made symbols and eroded the sense of what is natural or possible, the home page, the media event, and the IMAX theater offer intensified versions of the real. If the electrified landscape had begun the dematerialization of the built environment, the internet continued that process of abstraction, creating an infinitely malleable digital world. The next step is virtual reality.

Current narratives emphasize new figures in this digitized landscape, made possible by nomadic objects: modems, cellular phones, lap-top computers, and walkmen. The migrating executive is located nowhere in space but always available at an e-mail address. The home computer worker avoids the morning rush hour because the work is sent electronically to the house. The off-shore investment analyst sitting on a yacht or beach no longer needs to be anywhere near the stock exchanges or commodity markets to follow their movements. The computer hacker can illegally enter the business records of firms thousands of miles away. Such figures are not rooted anywhere. They have used technology to loosen their ties to place. Keeping them in mind, what was once called the settlement of the United States begins to appear as a continual unsettling and uprooting. The journey into American space leads not to Wright Morris's *Home Place* but to more restless movement. The electrification of the countryside does not stabilize its population, but unsettles it.

In the fifth, satiric narrative technologies have unintended outcomes. These satiric stories are about the inadequacy of the visions of the world represented in the other narratives.[21] For example, western irrigation led not to the spread of family farms into the desert, but rather to concentrated economic power in agribusiness. Building railroads and roads to national parks led not only to greater public appreciation, but to a flood of tourism that threatened the integrity of the sites. Electrification of the countryside, meant to revivify rural life, helped eliminate many family farms and later fostered the exodus of upper-middle-class people from suburbia. The 1939 world's fair in New York, originally presented as "The World of Tomorrow" becomes a nostalgic reminder the Depression era. The computer, imagined as a management's supreme instrument of control over labor, enabling centralization and reinforcing corporate hierarchies, turned out to have just the opposite potentialities. The space program, initially presented as a literal region of the New Frontier, swiftly lost its popularity as the public realized that despite its high cost, no homesteaders were going to the moon.

The apocalyptic scenario is an extreme variation on the narrative of unintended consequences. In it the outcome is so destructive that satire is no longer possible. For example, excessive pollution can lead to global warming, desertification, rising seas, and world-wide flooding. Likewise, as discussed in the chapter on energy narratives, during the 1970s fuel shortages suggested scenarios of political and economic collapse. Other examples include fears of nuclear annihilation, widespread sterility caused by excessive use of plastics and chemicals in food production, or life-threatening mutants produced by genetic engineering. In all such stories, human beings prove unable to control the technologies they unleash.

Each of these stories can be used to explain a range of events, but each finally fails as a master narrative. The first is invalid because it simply assumes that technology will be there as needed, without attending to the production of machines or the social definition of their use. The second story, in which machines aid geographical expansion and economic growth is essentially Whig history. Not only can technologies all too easily be understood as abstract forces which dictate change, but this account over-looks the dangers of centralized political and social power that technolo-gies make possible. In contrast, the third narrative overemphasizes this centralization, and lends itself to descriptions of extreme opression and class domination. The fourth narrative, in which new technologies compress space and time, generalizes from technical changes to psycho-logical effects. Beloved of Gingrich and the Tofflers, it suggests an auto-matic process and overlooks conflicting uses of machines in actual practice. The fifth, satiric narrative emphasizes the vanity of technological wishes and the ironies of history, but it can easily be overdone. In practice, not all technologies have unintended consequences. Some things do work as planned (e.g. Hoover Dam or Apollo 11). The sixth, apocalyptic narra-tive, has repeatedly warned Americans of the end of the world or of life "as we know it" due to any number of technological causes. This tends to be a one-dimensional theory of unavoidable catastrophe. With the notable exception of Native-American history, it is not of much use in looking at the past, since the United States has survived, although this story will always have its advocates as a narrative of the future.

Each of these six accounts may seem satisfactory in dealing with a particular case, and each can appeal to readers. But none of them is defin-itive. Identifying conflicting narratives facilitates exploration of the points of view of historical participants, who usually see the events of their lives as parts of larger stories. Contemplating a set of rival discourses also frees us from being the prisoner of any one account. Ultimately, particular narratives about technology and the construction of space are transient, as

Americans continually try to make sense of their world, modifying narratives to incorporate new experience. When events seem beyond their control, technology easily becomes the modern word for fate. Yet no machine is inevitable. Its configuration and specifications are the result of a myriad human decisions. Its public acceptance is not automatic, as every marketing department knows. Its uses are not preordained, and often prove surprising. Ironically, at the moment that government and corporations have institutionalized research and development with predetermined goals such as landing a man on the moon, at the moment when new consumer products are the result of immense investment and careful calculation, Americans seem to have a predilection for technological determinism. Yet their contradictory versions of this very idea suggest that technologies are not autonomous. Machines do not impinge on an abstract space, after arriving from some other sphere, but rather are created and given their diverse meanings within American culture's narratives and spaces.

NOTES

Introduction

[1] Merritt Roe Smith and Leo Marx, *Does Technology Drive History? The Dilemma of Technological Determinism* (Cambridge: MIT Press, 1994).

[2] Leo Marx, *The Emergence of a Hazardous Concept*, paper, New School for Social Research, delivered at *Technology and the Rest of Culture*, January, 1977.

[3] Emile Durkheim, *The Elementary Forms of the Religious Life*, in W. S. G. Pickering, ed., *Durkheim on Religion* (London: Routledge & Keegan Paul, 1975), 154.

[4] The classic work is Henry Nash Smith, *Virgin Land* (Cambridge: Harvard University Press, 1950), while Annette Kolody contributed to a revision of this work with her *The Land Before Her* (Chapel Hill: University of North Carolina Press, 1984).

[5] Burton Benedict, *The Anthropology of World's Fairs* (London: Scholar Press, 1983), p. 7.

[6] Newspaper article cited in Carl David Arfwedson, *The United States and Canada in 1832, 1833, and 1834* (New York: Johnson Reprint Corporation, 1969), Vol. 2, pp. 2–3.

[7] Michael Chevalier, *Society, Manners, and Politics in the United States* (New York: Anchor, 1961), pp. 296–297.

[8] Theodore Dreiser, *Sister Carrie* (New York: Viking/Penguin, 1981), pp. 126–127.

[9] J. B. Jackson, *Discovering the Vernacular Landscape* (New Haven: Yale University Press, 1984), p. 8.

[10] This moral conception of landscape is at the heart of several articles edited by Michael Sorkin, *Variations on a Theme Park: The New American City and the End of Public Space* (New York: Noonday, 1992).

[11] Gary Wills, *Inventing America* (New York: Doubleday, 1978).

[12] Hayden White, *Tropics of Discourse* (Baltimore: Johns Hopkins, 1978), p. 89.

[13] David Harvey, *The Condition of Postmodernity* (Oxford: Basil Blackwell, 1989), p. 350.

12 Hayden White, *Metahistory: The Historical Imagination in Nineteenth Century Europe* (Baltimore: Johns Hopkins University Press, 1973), pp. 29–31. See also Hayden White, *The Content of the Form* (Baltimore: Johns Hopkins University Press, 1987), pp. 26–57.

Chapter 1

1   For an exhaustive anthology of early descriptions of Niagara Falls, see Charles Mason Dow, *Anthology and Bibliography of Niagara Falls*, 2 vols. (Albany: State of New York, 1921). On the Grand Canyon, no such comprehensive work exists, but classic descriptions may be found in Frank Waters, *The Colorado* (New York: Rinehart and Company, 1946), and Joseph Wood Krutch, *The Grand Canyon* (New York: William Sloane Associates, 1958).

2   Henry James, *Aspern Papers, The Novels and Tales of Henry James*, vol. 12 (New York: Scribner's, 1961), p. 49.

3   In playing with the literary cliché of American innocence and American experience, Twain undermined the pretensions of the grand tour even as he acknowledged its status and desirability. Mark Twain, *The Innocents Abroad* (New York: Harper, 1911).

4   The brevity of American vacations is not a function of class. Blue- and white-collar workers have virtually the same vacation time as professionals, generally two weeks. Civil service workers and teachers have more time.

5   John Sears, *Sacred Places: American Tourist Attractions in the Nineteenth Century* (New York: Oxford University Press, 1989).

6   Canals were also important in early tourism. The Erie Canal, completed in 1825, made it possible to travel by water from New York to Buffalo and Niagara Falls. Indeed, the 500-kilometer-long Canal itself was considered to be an interesting tourist site. Immigrants and travelers to the West regularly traveled on the Erie during the 1830s and 1840s. See Carl David Arfwedson, *The United States and Canada in 1832, 1833, and 1834* (New York: Johnson Reprint Corporation, 1969), Vol. 2, pp. 277–87.

7   On railway competition between cities, see Zane L. Miller, *Urbanization of Modern America* (San Diego and New York: Harcourt Brace, Jovanovich, 1987), pp. 32–35. For more detail on railway competition and cooperation, see Alfred D. Chandler, *The Visible Hand* (Cambridge: Harvard University Press, 1977), pp. 122–144. On trackage in the United States and Europe, see Richard B. Morris, *Encyclopedia of American History* (New York: Harper and Row, 1970), p. 448.

8   On photographers and painters representing the American landscape, see Albert Boime, *The Magisterial Gaze: Manifest Destiny and American Landscape Painting, c. 1830–1865* (Washington: Smithsonian Institution, 1991), pp. 127–137.

9   As early as 1848 Boston alone had seven major terminals and more than 200 trains a week. Miller, p. 50.

[10] John Stilgoe, *Metropolitan Corridor* (New Haven: Yale University Press, 1983), p. 250.

[11] Ibid., p. 253.

[12] Elizabeth McKinsey, *Niagara Falls: Icon of the American Sublime* (Cambridge: Cambridge University Press, 1985).

[13] Dow, pp. 1132–1133. Niagara developed first as a private site, but in the late nineteenth century it was taken over by the State of New York and the Canadian Government,

[14] Dean MacCannell, *The Tourist: A New Theory of the Leisure Class* (New York: Schocken Books, 1976), pp. 118–125.

[15] McKinsey, p. 152.

[16] Sarah Margaret Fuller Ossoli, *Summer on the Lakes, in 1843* (Boston: Little, Brown, 1844), pp. 1–13. See also McKinsey, pp. 192–200.

[17] Stanford E. Demars, *The Tourist in Yosemite, 1855–1985* (Salt Lake City: University of Utah Press, 1991).

[18] On the photographic representation of the West, see Barbara Novak, *Nature and Culture* (New York: Oxford University Press, 1980). Nancy K. Anderson, "The Kiss of Enterprise: The Western Landscape as Symbol and Resource," in William H. Truettner, *The West as America: Reinterpreting Images of the Frontier, 1820–1920* (Washington: Smithsonian Institution, 1991), pp. 243–246.

[19] On photographic work for the Union Pacific Railroad and its use in lecturing, see Susan Danly, "Andrew Joseph Russell's The Great West Illustrated," in Susan Danly and Leo Marx, *The Railroad in American Art: Representations of Technological Change* (Cambridge: MIT Press, 1988), p. 100.

[20] Henry Hussey Vivian, *Notes of a Tour in America* (London: E. Stanford, 1878).

[21] For the history of early tourism in the West, see Earl Pomeroy, *In Search of the Golden West* (New York: Alfred Knopf, 1957). Prices in Anne Farrar Hyde, *An American Vision: Far Western Landscape and American Culture* (New York: New York University Press, 1990), p. 108.

[22] The standard history is Alfred Runte, *National Parks: The American Experience* (Lincoln: University of Nebraska Press, 1979).

[23] United States Department of the Interior, *Transportation Study: South Rim Village* (Grand Canyon National Park, July, 1990). Copy in the Grand Canyon Visitor Center and Museum Library.

[24] Characteristic questions about the Grand Canyon were compiled by park interpretive staff and posted at the Visitor's Center, where I copied them down in December, 1993.

[25] John Hance, *Personal Impressions of the Grand Canyon of the Colorado River* (San Francisco: Whitaker and Ray, 1899), pp. 62–63.

[26] US Department of the Interior, *Transportation Study*, p. 24.

[27] On the economic relationship between street traction companies and amusement parks, see David E. Nye, *Electrifying America: Social Meanings of a New Technology* (Cambridge: MIT Press, 1990), pp. 122–129.

28 The Niagara attraction was designed by Joseph Turner of New York and erected at Steeplechase Park at Coney Island. *Street Railway Journal*, 11 March, 1905, p. 481. [Copy, Smithsonian Museum of American History.]

29 John Urry, *The Tourist Gaze* (London: Sage, 1990).

Chapter 2

1 Turner first presented his thesis at the meeting of the American Historical Association at the Chicago World Fair. A summary of the defects of the Turner Thesis had become standard in textbooks by the 1950s. For an overview, see George Rogers Taylor, ed., *The Turner Thesis* (Boston: D. C. Heath, 1956). For further discussion see Michael P. Malone, "The Historiography of the American West," & David W. Noble, "Frederick Jackson Turner and Henry Nash Smith Revisited, 1890–1950–1990," in Rob Kroes, ed. *The American West As Seen by Europeans and Americans* (Amsterdam: Free University Press, 1989), pp. 1–18, 19–36.

2 See Richard White, "Frederick Jackson Turner and Buffalo Bill" in James R Grossman, ed., *The Frontier in American Culture* (Berkeley: University of California Press, 1994), p. 9.

3 Frederick Jackson Turner, *Frontier and Section* (Englewood Cliffs: Prentice-Hall, 1961), pp. 37–40.

4 For a summary, see Patricia Nelson Limerick, "The Adventures of the Frontier in the Twentieth Century," in Grossman, ed., pp. 67–95.

5 Gunther Barth, *Instant Cities: Urbanization and the Rise of San Francisco and Denver* (New York: Oxford University Press, 1975).

6 Joseph R. Conlin, "Grub and Chow: Food, Foodways, Class and Occupation on the Western Frontier," in Kroes, pp. 128–138.

7 Cited in Charles M. Coleman, *P.G. & E of California: The Centennial Story of Pacific Gas and Electric, 1852–1952* (New York: McGraw Hill, 1952), p. 56.

8 Letter of John Kreusi, January 12, 1892, Edison Pioneer Papers, Henry Ford Museum Library. Also see Sprague Electric Company, "Electric Transmission of Power for Mining Work" [pocket sized pamphlet], c. 1889, Henry Ford Museum Library.

9 General Electric Company, Electric Mine Locomotives, [catalogue] February, 1904; John Winthrop Hammond, *Men and Volts: The Story of General Electric* (Philadelphia: Lippincott, 1941), p. 91.

10 Coleman, p. 52.

11 Ibid., p. 53.

12 The sale to Ladd was noted in Edison Illuminating Company, Bulletins, 1882, p. 35, copy in Henry Ford Museum Library. The tower form of lighting used in Denver was installed in many cities by the Fort Wayne "Jenny" Electric Light Company, whose products are described in its advertising, National Museum of American History, Warshaw Collection, Electricity Series, Box 15.

13 Roger Sale, *Seattle: Past to Present* (Seattle: Washington University Press, 1976), pp. 72–3.

14 Thomas P. Hughes, *Networks of Power: Electrification in Western Society, 1880–1930* (Baltimore: Johns Hopkins University Press, 1983), pp. 270–281. On PG & E, see Walton Bean, *California: An Interpretative History* (New York: McGraw Hill, 1973), pp. 402–405.

15 M. Luckiesh, "Residence Lighting-Analysis of Home Lighting Contest Primers," National Electric Light Association, *Proceedings of the Forty-Eighth Convention, 1925*, pp. 715–718.

16 Richard Rudolph and Scott Ridley, *Power Struggle* (New York: Harper and Row, 1986), p. 45.

17 Samuel P. Hays, *Conservation and the Gospel of Efficiency* (New York: Athencum, 1979), pp. 5–26, 114–121. For the political history, see Phillip J. Funigiello, *Toward A National Power Policy: The New Deal and the Electric Utility Industry, 1933–1941* (Pittsburgh: University of Pittsburgh Press, 1973), pp. 3–20. For popular reaction, see "Politics Discovers a Power Trust," *Literary Digest*, 2 April 1927, p. 1027.

18 Robert Engler, ed., *America's Energy: Reports from The Nation* (New York: Pantheon Books, 1980), pp. 81–90.

19 Bean, p. 405.

20 Sprague Electric Company, "Electric Transmission of Power for Mill Work" (pocket sized pamphlet), c. 1889; copy in Henry Ford Museum Library. The standard work is Richard B. Duboff, *Electric Power in American Manufacturing, 1889–1958* (New York: Arno Press, 1979).

21 About 8–10 percent of the current was lost in the conversion of water power into electrical energy, another 9 percent in the transmission lines, and 8 percent more in the reconversion of energy in the motor. Sprague found that his system was 72 percent efficient. Ibid.

22 Hughes, pp. 269–270.

23 See Anthony Wallace, *Rockdale* (New York: Norton, 1978).

24 The West as a whole has 27 persons per square mile, compared to 306 in the Northeast, 79 in the Middle West, and 92 in the South. But this population is concentrated in cities: 84 percent of the West's population is urban, versus 79 percent in the Northeast, 71 percent in the Middle West, and 67 percent in the South. *Statistical Abstract of the United States, 1986* (Washington DC: United States Government Printing Office, 1985), pp. 10.

25 The standard work is Samuel Bass Warner, *Streetcar Suburbs* (Cambridge: Harvard University Press, 1962).

26 Barth, pp. 222–224.

27 US Census Bureau, *Special Report, Electric Industries, 1902* (Washington DC: Government Printing Office, 1906), p. 24.

28 Bean, p. 282; Roger Sale, pp. 61–62, 72.

29 Ibid., pp. 17–19.

30 George W. Hilton & John F. Due, *The Electric Interurban Railways in America* (Stanford: Stanford University Press, 1960); Bernard Meiklejohn, "Electricity Transforming Traffic," *World's Work*, Vol. 10, 1905, p. 6183.

31 Census, p. 230.

32 Brochure, "Orange Empire Trolley Trip," Copy in National Museum of American History, Warshaw Collection, Street Cars, Box 2.

33 "Terrible Accident at Tacoma," *Street Railway Journal*, 21 July 1900, p. 946.

34 Statistics from *People—Their Power: The Rural Electric Fact Book* (Washington DC: National Rural Electric Cooperative Association, 1980), p. 165.

35 Ednah Aiken, *The River* (Indianapolis: Bobbs-Merrill Co., 1914), p. 30.

36 Bean, p. 280. *Statistical Abstract*, pp. 640, 658.

37 Odie B. Faulk, *Arizona: A Short History* (Norman: University of Oklahoma Press, 1970), pp. 224–229. Troops were called out to prevent the construction of Parker Dam, later completed in 1941 and supplying water to Los Angeles.

38 Under this agreement, California was to get 4.4 million acre feet of water; Arizona 2.8 million, and Nevada but 300,000. Lawsuits continued until the 1960s, when the Supreme Court found in Arizona's favor.

39 Donald Worster, *Rivers of Empire: Water, Aridity, and the Growth of the American West* (New York: Pantheon, 1985).

40 Woody Guthrie, "Do-Re-Mi," *Library of Congress Recordings* (CDs 1041, 1042, 1043) (Cambridge: Rounder Records, 1988).

41 Leo Marx, *The Machine in the Garden* (New York: Oxford University Press, 1964).

42 Timothy J. Healy, *Energy and Society* (San Francisco: Boyd & Fraser, 1976), p. 164.

43 William K. Wyant, *Westward in Eden: The Public Lands and the Conservation Movement* (Berkeley: University of California Press, 1982), pp. 255–6.

44 See John William Ward, *Andrew Jackson: Symbol for an Age* (New York: Oxford University Press, 1955). Roderick Nash, *Wilderness and the American Mind* (New Haven: Yale University Press, 1967).

45 John Sears, *Sacred Places: American Tourist Attractions in the Nineteenth Century* (New York: Oxford University Press, 1989), pp. 122–65. Alfred Runte, *National Parks: The American Experience* (Lincoln: Nebraska University Press, 1987). Hal Rothman, *America's National Monuments: The Politics of Preservation* (Lawrence: University Press of Kansas, 1994).

46 The federal government controls 86 percent of Alaska; eliminating it, however, Washington still has 48 percent of California; 64 percent of both Idaho and Utah; 44 percent of Arizona; 85 percent of Nevada, etc. Census Bureau, *Statistical . . .* p. 196.

47 Instead of the lights themselves being visible, fixtures were at ground level or disguised, so that the spectator did not notice them. By separating the light from the objects to be lighted, attention focused attention on the buildings. The sense of three-dimensionality was preserved by projecting dark red "luminous shadows" into corners and other areas where white floodlights cast shadows.

48 W. D. Ryan, "Illumination of the Panama–Pacific Exposition," *Scientific American Supplement*, 79 (12 June 1915), pp. 376–7; H. M. Wright, "Panama–Pacific Exposition at Night," *Scientific American* (24 April 1915); H. Whitaker, "In a Blaze of Glory," *Sunset* 34 (March 1915), pp. 511–18.

49 On world fairs, see Robert Rydell, *All the World's a Fair* (Chicago: University of Chicago Press, 1985).

50 Also see David E. Nye, "Social Class and the Electrical Sublime, 1880–1915," in Rob Kroes, ed., *High Brow Meets Low Brow* (Amsterdam: Free University Press, 1988), pp. 1–20.

51 Umberto Eco, *Travels in Hyperreality* (New York: Harcourt Brace Jovanovich, 1987), p. 44.

Chapter 3

1 J. B. Jackson, "The Westward Moving House," in Ervin Zube, ed., *Landscapes: Selected Writings of J.B. Jackson* (Amherst: University of Massachusetts Press, 1970), pp. 1–31.

2 On the mythology of the high plains, see Henry Nash Smith, *Virgin Land* (Cambridge: Harvard University Press, 1950). On the harsh realities, see Donald Worster, *Dust Bowl: The Southern Plains in the 1930s* (New York: Oxford University Press, 1979).

3 David B. Danbom, *Born in the Country: A History of Rural America* (Baltimore: Johns Hopkins University Press, 1995), pp. 161–230.

4 Wright Morris, *The Home Place* (Lincoln & London: University of Nebraska Press, 1968). [first published, 1948] Morris's photography was displayed in 1975 in a two-hundred-print retrospective, with a catalogue published as *Wright Morris: Structures and Artifacts, Photographs, 1933–1954* (Lincoln: University of Nebraska Press, 1975). Morris stopped making photographs in 1954, but has been an occasional critic: "In Our Image," *The Massachusetts Review* Vol XIX, No. 4, 1978, pp. 633–43; see also his *Earthly Delights, Unearthly Adornments* (New York: Harper and Row, 1978).

5 Erskine Caldwell and Margaret Bourke-White, *You Have Seen Their Faces* (New York: Modern Age Books, 1937). William Stott, *Documentary Expression and Thirties America* (New York: Oxford University Press, 1973), p. 221.

6 James Agee and Walker Evans, *Let Us Now Praise Famous Men* (New York: Ballantine Books, 1966, first published 1941). Stott, ibid., devotes two chapters to this work.

7 Wright Morris, *A Cloak of Light: Writing My Life* (New York: Harper and Row, 1985), pp. 51, 56–7.

8 James Laughlin, ed., *New Directions in Prose and Poetry, 1940* (Norfolk Conn.: New Directions, 1940). Wright Morris, *The Inhabitants* (New York: Charles Scribner's Sons, 1945).

9 John Steinbeck, *The Grapes of Wrath* (New York: The Viking Press, 1958, first published in 1939), p. 107. *Home Place*, pp. 26–9.

[10] Morris, *Home Place*, frontispiece.

[11] *New York Times*, 4 July, VI, p. 14. 1948. Reviewers were on the whole positive about Morris's first book, but readers were confused, unable to decide if it was a photography book or a work of fiction. Sales were poor, and when the author delivered his second photo-novel, *A World in the Attic*, to Scribner's, they bluntly told him that they would be glad to publish it without images, but not as another photo-text volume. Morris, *A Cloak of Light*, pp. 144–5.

[12] A. D. Coleman, *Light Readings* (New York: Oxford University Press, 1979), p. 244.

[13] Fredric Jameson, *Marxism and Form* (Princeton: Princeton University Press, 1971), p. 355.

[14] Morris, *The Home Place*, photographs opposite pp. 1, 51, 128, 177; quotation, pp. 35–7.

[15] Ibid., p. 26.

[16] Cited in Stott, p. 219.

[17] Morris, *The Home Place*, p. 132.

[18] Ibid., pp. 41, 176.

[19] Ibid., p. 176.

[20] Cited in Joseph J. Wydeven, "Photography and Privacy: The Protests of Wright Morris and James Agee," *The Midwest Quarterly*, Vol XXIII (Autumn, 1981), p. 106.

[21] Morris, *The Home Place*, p. 141.

[22] Ibid., p. 135.

[23] On the camera's inherent lack of realism, see David E. Nye, *Image Worlds* (Cambridge: MIT Press, 1985), pp. 31–56.

[24] Morris, *The Home Place*, pp. 37–8.

[25] Simon Watney, "Making Strange: The Shattered Mirror," in Victor Burgin, ed., *Thinking Photography* (London: Macmillan, 1982), p. 173.

[26] In 1887 the Montgomery Ward catalogue contained 24,000 items. The majority of its business was in rural areas. On the mail-order house's penetration of rural markets in the nineteenth century see Alfred D. Chandler, *The Visible Hand* (Cambridge: Harvard University Press, 1977), p. 230.

[27] Wright Morris, *God's Country and My People* (New York: Harper and Row, 1968).

[28] On the ficticity of realism, see Jacques Derrida, *Of Grammatology* (Baltimore and London: Johns Hopkins University Press, 1976); and Roland Barthes, *S/Z* (New York: Hill and Wang, 1972). Barthes had trouble reconciling his theories of language with photography, see "The Photographic Message," in his *Image, Music, Text* (New York: Hill and Wang, 1977).

[29] Morris, *The Home Place*, p. 151.

[30] Ibid., pp. 154–155.

Chapter 4

1 For a survey of the electrification of Middletown from the 1880s until World War II, see David E. Nye, *Electrifying America: Social Meanings of a New Technology* (Cambridge: MIT Press, 1990), chapter 1.

2 My purpose here is not to follow Hayden White's typology, but I must acknowledge a debt to his *Metahistory* (Baltimore: Johns Hopkins University Press, 1973). My thoughts on narrative explanations and energy are worked out in the chapter five, "Energy Narratives."

3 Paul K. Conkin, *Tomorrow a New World* (Ithaca, New York: Cornell University Press, 1959), p. 11.

4 Robert S. and Helen M. Lynd, *Middletown in Transition: A Study in Cultural Conflicts* (New York: Harcourt Brace Jovanovich, 1982), pp. 8, 53.

5 Ralph Borsodi, *This Ugly Civilization* (New York: Simon and Schuster, 1929); *Flight from the City* (New York: Harper & Brothers, 1933).

6 Paul K. Conkin, *The Southern Agrarians* (Knoxville: University of Tennessee Press, 1988), p. 103. See Ransom's essay "What does the South Want?" in Herbert Agar and Allen Tate, eds., *Who Owns America? A New Declaration of Independence* (Boston: Houghton Mifflin, 1936), pp. 178–193.

7 Derived from the urban theory of the nineteenth-century Spanish architect Soria y Mata, Broadacre would be a linear strip that could be infinitely extended when town–country barriers were broken down. Giorgio Cucci, "The City in Agrarian Ideology and Frank Lloyd Wright," in *The American City from the Civil War to the New Deal* (London: Granada, 1980), pp. 342–3, 360–2.

8 The three green-belt communities were near Washington, Milwaukee, and Cincinnati. For a brief discussion of de-urbanization in the New Deal, see Zane L. Miller and Patricia M. Melvin, *The Urbanization of Modern America* (New York: Harcourt Brace Jovanovich, 1987), pp. 169–172. "The City," 55 minutes, 1939, distributed by Museum of Modern Art, New York.

9 *Statistical Abstract of the United States*, 1986 and 1992.

Year	Farm population as a % of total population.
1950	15.3
1955	11.6
1960	8.7
1965	6.4
1970	4.8
1975	4.2
1980	2.7
1985	2.3
1990	1.9

10 Martin T. Cadwallader, *Analytical Urban Geography: Spacial Patterns and Theories* (Englewood Cliffs: Prentice-Hall, 1985), p. 104.

[11] See John M. Wardwell, "The Reversal of Nonmetropolitan Migration Loss," in *Rural Society in the US: Issues for the 1980s*, Don A. Dillman and Daryl J. Hobbs eds. (Boulder: Westview Press, 1982), pp. 23–33.

[12] Gallup Opinion Index, survey of adults. 1985. Preferences were for the city, 13 percent; the suburbs, 25 percent; small towns 36 percent; and farm or rural, 25 percent. The desire to live in cities has fluctuated, but has not been over 20 percent since 1966. Polls cited in Daniel J. Elazar, *Building Cities in America* (Lanham, Maryland: Hamilton Press, 1987). William Stephens, "A Rural Landscape but an Urban Boom," *New York Times*, 8 Aug. 1988, p. 1.

[13] Alfred D. Chandler, "The Beginnings of 'Big Business' in American Industry." Reprinted in Richard Tedlow and Richard John, eds., *Managing Big Business* (Boston: Harvard Business School Press, 1986). Alfred D. Chandler, *The Visible Hand* (Cambridge: Harvard University Press, 1977). Alfred Chandler and Richard Tedlow, *The Coming of Managerial Capitalism* (Homewood, Illinois: Richard D. Irwin, 1985). On farm electrical sales, see Henry Ford Museum, Greenfield Village, trade catalog collection, "The Electro-Set Handybook and Catalogue," 1916.

[14] Michael A. Bernstein, *The Great Depression: Delayed Recovery and Economic Change in America, 1929–1939* (Cambridge: Cambridge University Press, 1987), pp. 19–20, 32–36.

[15] Lynd, *Middletown in Transition*, pp. 52–53.

[16] "Indiana & Michigan Electric Company, Its Origin and Development," Indiana–Michigan Company Archives, Muncie, Indiana, p. 27. For early history, see Alexander R. Holliway, "Rural Electrical Development in Indiana." (Indianapolis: 1914). For the Indiana REA, see Emily Born, *Power to the People: A History of Rural Electrification in Indiana* (Indianapolis: Indiana Statewide Association of Rural Electric Cooperatives, 1985).

[17] *Muncie Morning Star*, 26 April 1935.

[18] Marquis Childs, *The Farmer Takes a Hand: The Rural Electric Power Revolution in Rural America* (New York: Doubleday, 1952), and D. Clayton Brown, *Electricity for Rural America: The Fight for the REA* (Westport: Greenwood Press, 1980). Also see his "Farm Life: Before and After Electrification," *Proceedings of the Annual Meeting, the Association for Living Historical Farms and Agricultural Museums* (Washington DC: Smithsonian Institution, 1975).

[19] Preston J. Hubbard, *Origins of the TVA: The Muscle Shoals Controversy, 1920–1932* (New York: Norton, 1961). On FDR's interest in TVA see Frank Friedel, *Franklin D. Roosevelt: Launching the New Deal* (Boston: Little Brown, 1973), p. 353. For a short account of the TVA's early years see Joseph G. Knapp, *Advance of American Cooperative Enterprise, 1920–1945* (Danville, Illinois: Interstate Printers, 1973), chapter XVI. For more detail, Thomas K. McCraw, *TVA and the Power Fight, 1933–1939* (Philadelphia: Lippincott, 1971).

[20] By 1979 the REA generated 36 percent of its own energy, with little reliance on nuclear plants. It purchased the rest from public agencies such as the TVA (36 percent) and private utilities (28 percent).

21 Hawaii, Massachusetts, Connecticut, and Rhode Island were the only states that never participated in the REA. Three were densely populated states where private utilities and streetcar companies had managed to reach most farms, while Hawaii was still a territory in the 1930s.

22 See Roy Talbert, *FDR's Utopian: Arthur Morgan of the TVA* (Oxford, Mississippi: University of Mississippi, 1987).

23 For more on the conflict between the REA and the utilities, see Childs, *The Farmer Takes a Hand*, pp. 50–87.

24 Files of Henry Slattery, Box 3, Record Group 221, National Archives.

25 George H. Foster, "A Study of the Rural Electrification Project in Boone County, Indiana" BA thesis, Purdue University, June 1939, p. 9.

26 See *Electrifying America*, pp. 318–9.

27 The REA Engineering Division estimated that "the operating life of refrigerators now on the market is about 8 years, while upkeep cost amounts to about $1.50 a year." Taking depreciation into account, the annual cost of a refrigerator was about $25. "Electric Household Refrigerators," Office of Henry Slattery, 1939–1945, Box 2, Record Group 221, REA, National Archives. For more examples, see "Henry County, Load Building Campaign Data," Appliance Survey, 15 Sept. 1939, Rural Electrification Administration, Record Group 221, Office of Harry Slattery, 1939–1945, National Archives. Also see "More Appliance Buying," *Rural Lines* (the REA magazine), 1 Jan. 1939.

28 Rural Electrification Administration, Record Group 221, Office of Harry Slattery, 1939–1945, National Archives.

29 Richard A. Pence, ed., *The Next Greatest Thing* (Washington: NRECA, 1984), p. 100.

30 See David E. Nye, "Redeeming City and Farm: The Iconography of New Deal Photography," in Steven Ickringill, ed. *Looking Inward, Looking Outward: From the 1930s through the 1940s* (Amsterdam: Free University Press, 1990).

31 Paul K. Conkin, ed., *TVA: Fifty Years of Grassroots Bureaucracy* (Urbana: University of Illinois Press, 1983). Eleanor Buckles, *Valley of Power*, (New York: Creative Age Press, 1945), p. 114.

32 The classic work from this point of view is Philip Selzick, *TVA and the Grass Roots: A Study in the Sociology of Formal Organization* (Berkeley: University of California Press, 1949).

33 David E. Whisnant, *Modernizing the Mountaineer: People, Power, and Planning in Appalachia* (Knoxville: University of Tennessee Press, 1994), p. 61.

34 Ibid., p. 220.

35 Harvey, p. 350. See discussion in the Introduction.

Chapter 5

[1]   One caveat must be stated at the outset. I will discuss only male energy narratives, which characteristically concern power systems and their problems, growth, limits, success, and failure. A corresponding study of women's energy narratives might focus on biological rather than mechanical metaphors. That essay would explore such topics as generativity and barrenness, exclusion and empowerment. I am concerned with the narrative designs fashioned by white men, who have long dominated American politics and culture.

[2]   William Wister Haines, *High Tension* (Boston: Little, Brown, 1938). Such reform novels were an important part of American culture in the early twentieth century. They were brought out by respected houses, including Bobbs-Merrill, Macmillan, Grosset & Dunlap; and Little, Brown, and they played a role in political life.

[3]   Louis C. Hunter, *A History of Industrial Power in the United States, 1780–1930*, Vol. 2, *Steam Power* (University Press of Virginia, 1985), p. 74.

[4]   Ibid., pp. 75–84.

[5]   Ronald E. Martin, *American Literature and the Universe of Force* (Durham: Duke University Press, 1981), pp. 27–29, and passim.

[6]   Henry Adams, *The Education of Henry Adams* (Boston: Houghton Mifflin, 1961), chapters 25 and 33.

[7]   A. J. Greimas, *Sémantique structurele* (Paris: Larousse, 1966); *Du Sens* (Paris: Seuil, 1970).

[8]   See Will Wright, *Sixguns and Society* (Berkeley: University of California, 1976); Janice Radway, *Reading the Romance* (Chapel Hill: University of North Carolina, 1984).

[9]   Northrop Frye, *Anatomy of Criticism* (Princeton: Princeton University Press, 1957). Hayden White, *Metahistory: The Historical Imagination in Nineteenth Century Europe* (Baltimore: Johns Hopkins University Press, 1973).

[10]   White, p. 10.

[11]   For discussion, see John F. Kasson, *Civilizing the Machine* (Harmondsworth: Penguin, 1977), pp. 178–9.

[12]   Roland Marchand, *Advertising the American Dream: Making Way for Modernity, 1920–1940* (Berkeley: University of California Press, 1985), p. 363.

[13]   Charles Ripley, *The Romance of Power* (New York: National Electric Light Association, 1928). The 91 pages of text were illustrated with the 104 of the images used in the slide show. Archives of the National Museum of American History, Warshaw Collection, Box 9, pp. 66–8, 91. The majority of the images used can be found in the General Electric Photographic Archives, Schenectady, New York. The role of such materials in public relations campaigns is discussed in chapter eight of David E. Nye, *Image Worlds: Corporate Identities at General Electric* (Cambridge: MIT Press, 1985).

[14]   Alexander Otis, *The Man and the Dragon* (Boston: Little, Brown and Company, 1910). Henry George, Jr., *The Romance of John Bainbridge* (New York: Macmillan Company, 1906).

15 See, for example, Roland Phillips, "Problem of Municipal Ownership," *Harper's Weekly* 51 (September 1907): 1037; Ernest E. Williams, "How London Loses by Municipal Ownership," *North American Review* 183 (October, 1906): pp. 729–736.

16 See Herman W. Chaplin, *The Coal Mines and the Public* (J. B. Millet Company, 1902).

17 *New York Times*, 7 Nov. 1974.

18 *New York Times*, 3 Dec. 1974. Duke University Library, Special Collections, J. Walter Thompson Corporate Archives, Competitive Advertisements B110–B150, 1974, Box 3 folder 11.

19 Eleanor Buckles, *Valley of Power* (New York: Creative Age Press. Second printing, 1945), p. 123.

20 Cecilia Tichi, *Shifting Gears: Technology, Literature, Culture in Modernist America* (Chapel Hill: University of North Carolina, 1987), pp. 117–131.

21 New York Board of Trade, *Business Speaks* (April 1974), p. 4.

22 On Henry Ford's energy views, see David E. Nye, *Henry Ford: Ignorant Idealist* (New York: Kennikat Press, 1979), chapter 4.

23 On Edison as hero, see David E. Nye, *The Invented Self: An Anti-biography of Thomas A. Edison* (Odense: Odense University Press, 1983).

24 *Collier's* and *The Saturday Evening Post* quotations from Stephen Hilgartner, Richard C. Bell, Rory O'Connor, *Nukespeak: The Selling of Nuclear Technology in America* (Harmondsworth: Penguin Books, 1982).

25 Robert L. Heilbroner, "Growth and Survival," *Dialogue*, Winter, 1973. [Reprint of paper published by the Council on Foreign Relations in 1972.]

26 Amory Lovins and L. Unter Lovins, *Brittle Power: Energy Strategy for National Security* (Andover, Mass.: Brick House Publishing, 1982), pp. 18–20.

27 Thomas A. Edison, *The Diary and Sundry Observations* (New York: Philosophical Library, 1948), pp. 231–232.

28 For example, see Matthew Luckiesh, *Artificial Sunlight: Combining Radiation for Health and Light for Vision* (New York: Van Nostrand Company, 1930). Edison's views of the social effects of the electric light in "Edison's Prophecy: A Duplex, Sleepless, Dinnerless World," *Literary Digest* (14 Nov. 1914), pp. 966–968.

29 Murray Melbin, *Night as Frontier: Colonizing the World After Dark* (New York: The Free Press, 1987, p. 127.

30 Mark Twain, *A Connecticut Yankee in King Arthur's Court* (New York: New American Library, 1980).

31 Henry Adams, *The Education of Henry Adams* (Boston: Houghton Mifflin, 1981), chapter 25. Also, Henry Adams, *The Degradation of Democratic Dogma* (New York: Harper Torchbook, 1969). For assessments of Adams, see Ernest Samuels, *Henry Adams: The Major Phase* (Cambridge: Harvard University Press, 1964), chapter 11 and Cecilia Tichi, *Shifting Gears* (Chapel Hill: University of North Carolina Press, 1987), pp. 152–164.

32 For an account of these and other monster films of the era, see Spencer R.

Weart, *Nuclear Fear: A History of Images* (Cambridge: Harvard University Press, 1988), pp. 191–2.

33 Ibid., p. 238.

34 Paul Theroux, *O-Zone* (Harmondsworth: Penguin, 1987). Martin, pp. 27–9, and passim.

35 Thomas Burton, (Stephen Longstreet, pseud.), *Bloodbird* (New York: Smith and Durrell, 1941).

36 The stock market crashes, undermining the utility's economic position and paralyzing work on the dam. Then the Winthrop's town burns to the ground, a victim of shoddy water pipes that freeze. The conflagration is presented as a natural event.

37 Thomas Jefferson, *Notes on the State of Virginia* (Gloucester: Peter Smith, 1976), pp. 156–158.

38 Ernest Callenbach, *Ecotopia: A Novel* (New York: Bantam, 1977).

39 E. F. Schumacher, *Small is Beautiful: Economics as if People Mattered* (Harper and Row, 1973).

40 Televised speech, 18 April 1977. Quoted in Jimmy Carter, *Keeping Faith* (London: Collins, 1982).

41 Ronald Reagan, Acceptance Speech, 17 July 1980. Transcript, *New York Times*, 18 July 1980. After the election his Secretary of the Interior permitted more drilling for oil and natural gas on federal lands. But before this energy reached the market, OPEC's efforts to control prices broke down, and the price of oil began to drop. By 1985 it was less than half what it had been in 1980.

42 There is another interesting category, which I might call narratives of transcendence, which are closely related to those of limitation. In this connection, see Allen Ginsberg's poem, "Plutonian Ode."

43 Donella H. Meadows, et al., *The Limits to Growth* (London: Pan Books, 1972). The book sold widely in the United States and had gone through five printings in England alone by 1979.

44 In part this framework has already been explored in Stephen Kern, *The Culture of Space and Time* (London: Weidenfeld and Nicolson, 1983).

45 Pierre Macherey, *A Theory of Literary Production* (London: Routledge, & Kegan Paul, 1978), p. 155.

46 F. Scott Fitzgerald, *The Great Gatsby* (Harmondsworth: Penguin, 1964). On electrification see pages 42, 45, 79, 108, 171.

47 Further background in David E. Nye, *Electrifying America: Social Meanings of a New Technology* (Cambridge: MIT Press, 1990), chapter 2, "The Great White Way," and chapter 5, "Flexible Factory."

Chapter 6

1 Hayden White, *Metahistory: the Historical Imagination in Nineteenth Century Europe* (Baltimore: Johns Hopkins University Press, 1973), Introduction.

2 E. L. Doctorow, *World's Fair* (New York: Fawcett Crest, 1985). All subsequent page references to this edition, abbreviated WF.

3 Hayden White, *Tropics of Discourse: Essays in Cultural Criticism* (Baltimore: Johns Hopkins University Press, 1978), p. 96.

4 None of the group he discusses makes up facts or intentionally employs anachronisms.

5 E. L. Doctorow, "False Documents," reprinted in Richard Trenner, ed., *E. L. Doctorow: Essays and Conversations* (Princeton, New Jersey: Ontario Review Press, 1983).

6 See David E. Nye, *Henry Ford: Ignorant Idealist* (New York: Kennikat Press, 1979), pp. 59–69.

7 Cushing Strout, "Historicizing Fiction and Fictionalizing History: The Case of E. L. Doctorow," *Prospects* 5, 1980, p. 428. Strout makes many of the same points in "The Veracious Imagination," *Partisan Review* 50, 1983, pp. 428–43.

8 Ibid. (*Prospects*), p. 435.

9 E. L. Doctorow, *Ragtime*. (London: Macmillan, 1976) p. 107. The time and cost of this fictional trip is confirmed in a guide published by the *Brooklyn Daily Eagle*, "Trolley Exploring Trips About New York," (New York: 1910). Copy, Smithsonian Museum, Warshaw Collection, Box 2.

10 The inscription: "A raree show is here/ With children gathered round. . ."

11 George Plimpton, "Interview with E. L. Doctorow," *Paris Review*, 1983:3, pp. 27, 30. [emphasis added]

12 Dean Flower criticized precisely this aspect of the novel: "Doctorow seems to be reconstructing his first decade not so much to understand himself or ascertain his identity. . . as to evoke the period, passing around a whole lot of old snapshots, setting it down for posterity and the sociologists. . . . For all its elegance and clarity, it's a document of the time period instead of the person." "Fables of Identity," *Hudson Review*, Spring, 1986, p. 311.

13 Plimpton, p. 34. Doctorow's *Lives of the Poets* (London: Michael Joseph, 1984) contains a story, "The Writer in the Family" in which a boy writes fictional letters that are meant to be taken as true.

14 WF, p. 371.

15 Plimpton, p. 27.

16 Robert Towers, "Three Part Inventions," *New York Review of Books*, 10 December 1985.

17 See White, *Metahistory*, pp. 59–60.

18 WF, p. 16.

19 White, *Metahistory*, p. 14.

20 Mark Twain, *Adventures of Huckleberry Finn*, in George McMichael, ed., *Anthology of American Literature*. (New York: Macmillan, 1980), pp. 440–441.

21 WF, p. 146.

22 WF, p. 363.

23 Towers, op. cit.

24 Paul Levine, *E. L. Doctorow* (London: Methuen, 1985), pp. 59–60.

25 Cited in Levine, p. 51.

26 "In Walter Benjamin's brilliant essay 'The Story Teller: Reflections on the Works of Nikolai Leskov,' I read that storytelling in the Middle Ages was primarily a means of giving counsel. . . . each story was honed by time and many tellers. If the story was good the counsel was valuable and therefore the story was true." Doctorow, "False Documents," p. 18.

27 The Frankfurt School had curious views about popular culture, tending toward belief in a golden age of art and storytelling in the pre-industrial period before mechanical reproduction turned art into a commodity. This theory has been rejected by such recent students of popular culture as Fredric Jameson, Janice Radway, and John Kasson, who instead of contrasting popular culture with the high culture of the past, have seen it as a set of open-ended myths used by the working classes as a way to escape from their everyday lives.

28 *New York World's Fair Official Guide* (New York: Exposition Publications, 1939) reveals no such act in the amusement area, but these attractions changed often and there were roughly similar acts, including "Salvador Dali's Living Pictures."

29 WF, pp. 365–366.

30 Bond subscribers listed in *New York Times*, 1 May 1939; p. 6; costs, 30 April 1939, p. 3.

31 For Washington memorials, see *New York Times*, Special Supplement on the World's Fair, 30 April 1939, p. 29. Inauguration re-enacted, *New York Times*, 1 May 1939, p. 8.

32 WF, p. 327

33 Ralph Waldo Emerson, "The American Scholar," in Brooks Atkinson, *Selected Writings of Emerson* (New York: Modern Library, 1950), p. 46.

34 Burton Benedict, *Anthropology of World's Fairs* (London: Scholar Press, 1983), p.5. At Chicago in 1893 were vast artificial lagoons and waterways; Buffalo in 1901 featured a giant man-made waterfall; and in 1915 San Francisco copied both the Grand Canyon and Yellowstone Park, complete with working geysers.

35 Ibid., p. 6.

36 *New York Times*, 30 April 1939, Special Supplement, pp. 8, 9.

37 WF, p. 325.

38 Daniel Boorstin, *The Americans: The Democratic Experience* (New York: Vintage, 1974), p. 148.

39 Cushing Strout, "Historicizing Fiction. . .", p. 435.

Chapter 7

1 André Berne Joffroy, *Zigzag parmi les personnages de La Fée Electricité* (Paris: Musée d'Art Moderne de la Ville de Paris, n.d.) I am indebted to Mick Gidley who first drew my attention to this painting.

2 This process occured both in France and the United States. See Rosalind Williams, *Dream Worlds: Mass Consumption in Late Nineteenth-Century France* (Berkeley: University of California Press, 1982), p. 59, and Roland Marchand, "Corporate Imagery and Popular Education: World's Fairs and Expositions in the United States, 1893–1940," in David E. Nye and Carl Pedersen, eds., *Consumption and American Culture*, (Amsterdam: Free University Press, 1991), pp. 18–33. Large fairs were held in Cincinnati (1883), New Orleans (1884), Chicago (1893), Atlanta (1895), Omaha (1898), Buffalo (1901), St. Louis (1904), Seattle (1909), San Francisco (1915) and San Diego (1915). Also see John Cawelti, "America on Display, 1876, 1893, 1933," in *America in the Age of Industrialism*, ed. Frederic C. Jaher (New York: Free Press, 1968), pp. 317–63.

3 Louis C. Hunter, *A History of Industrial Power in the United States, 1780–1930*. Vol. 2, *Steam Power* (Charlottesville: University Press of Virginia, 1985), pp. 295–99.

4 Alain Beltran and Patrice A. Carré, *La fée et la servante: La société française face à l'électricité* (Paris: Editions Belin, 1991), pp. 64–72.

5 Siemens had exhibited a smaller electric tram in 1879 at the electrical fair in Berlin, but this did not carry the general public.

6 J. P. Barrett, *Electricity at the Columbian Exposition* (Chicago: R. R. Donnelley & Sons, 1894), pp. 16–19.

7 Harold C. Passer, *The Electrical Manufacturers, 1875–1900* (Cambridge: Harvard University Press, 1953), p. 94.

8 "Historical Sketch of the Foreign Business of the General Electric Company," *General Electric Digest*, Vol 2: 4, (1922), pp. 5–8; No. 5, pp. 15–19; No. 6, pp. 4–9.

9 Office address list inside front cover, *General Electric Digest* (Dec. 1929), Vol. 9:6.

10 General Electric, "Facts About General Electric Company of Interest to Stockholders." 25 July 1932. Copy in General Electric Corporation, Schenectady Library.

11 *General Electric Digest*, Vol 2:5, 1922, pp 18–19.

12 As late as 1940 it had 21 percent of Osram, 34 percent of the French Compagnie des Lampes, 12 percent of Phillips, and 40 percent of AEG. I. Arthur A. Bright, Jr., *The Electric Lamp Industry* (New York: Macmillan, 1949), p. 309.

13 *Louisville Courier-Journal*, 4 July 1883. Hammond Papers, General Electric Library, Schenectady, L 5410–5414.

14 Jane Mark Gibson, The International Electrical Exhibition of 1884 and the National Conference of Electricians" MA Thesis, University of Pennsylvania, 1984, p. 44 [fountain], p. 85 [Edison].

15 *Journal of the Franklin Institute*, May 1886, p. 121.

16 *Cincinnati Commercial Gazette*, 10 June 1888, p. 4. Copy, National Museum of American History, Archives, Hammer Papers, Series 3, Box 1.

17 Williams, p. 84.

18 A. F. Dickerson, "Spectacular Lighting," *Proceedings, National Electric Light Association Convention* 47 (1924), p. 485.

19 The same ideas, updated, were used in New York in 1939.

20 From the caption accompanying a full-page reproduction in *The Columbian Gallery: A Portfolio of Photographs from the World's Fair* (Chicago: The Werner Company, 1894), p. 86.

21 J. C. Levinson, et al. eds., *The Letters of Henry Adams* (Cambridge: Harvard University Press, Vol. 4, 1988), p. 132.

22 Theodore Dreiser, *Newspaper Days* (Philadelphia: University of Pennsylvania Press, 1991), pp. 309–10.

23 "Shown by Electric Light," *Omaha Daily Bee*, 2 June 1898, p. 1. I want to thank David Rotterman, KYNE-TV who drew my attention to this article and other local materials.

24 Cited in James B. Haynes, *History of the Trans-Mississippi and International Exposition of 1898* (Omaha: privately printed, 1910), p. 141. See also John A. Wakefield, "A History of the Trans-Mississippi and International Exposition," 20 May 1903, (typescript) Omaha Public Library.

25 Quotation in Williams, p. 85.

26 See Richard D. Mandell, *Paris, 1900* (Toronto: University of Toronto Press, 1967), pp. 112–3.

27 For a more detailed account, see David E. Nye, *Electrifying America: Social Meanings of a New Technology* (Cambridge: MIT Press, 1990), pp. 43–7.

28 Marshall Everett, *The Book of the Fair, The Greatest Exposition the World Has Ever Seen. A Panorama of the St. Louis Exposition* (Henry Neil: St. Louis, 1904), pp. 201–3.

29 On the Hudson-Fulton Celebration, see David E. Nye, "Republicanism and the Electrical Sublime," *American Transcendental Quarterly*, Fall, 1990, pp. 185–199.

30 Ben Macomber, *The Jewel City* (John H. Williams: San Francisco and Tacoma, 1915), chapter XIII, "The Exhibition Illuminated."

31 W. D'A. Ryan, "Illumination at the Panama–Pacific Exposition," *Scientific American*, supplement 2056, Vol 79–80, p. 377. Ryan, a General Electric employee, was chief of illumination at the exposition.

32 Dreiser, p. 310.

33 Century of Progress, *Official Guide: Book of the Fair* (Chicago, 1933), p. 23. The traditional light bulb produced so much heat that it was unsuited to a large windowless building of rectangular design unless it had good air-conditioning or powerful ventilators. Fluorescent bulbs would solve this problem, but not until the end of the 1930s.

34 Ibid., p. 29.

35 "Proposed Basis of Illumination at the New York World's Fair," from *Edison Life* (n.d.) courtesy of Boston Edison Company.

36 Wentworth M. Potter and Phelps Meaker, "Luminous Architectural Elements," Cleveland: General Electric Company, NELA Park Engineering, Dec. 1931. Copy in Hammond Papers, General Electric Library, Schenectady, W4061.

37 "Proposed Basis of Illumination at the New York World's Fair," from *Edison Life*. n.d. courtesy of Boston Edison Company. See Marchand, in David E. Nye and Carl Pedersen, eds., pp. 21–3.

38 Helen A. Harrison, "The Fair Perceived: Color and Light as Elements in Design and Planning," in Harrison, ed., *Dawn of a New Day: The New York World's Fair, 1939/40* (New York: New York University Press, 1980), pp. 43–56.

39 *Official Guide*, p. 168.

40 For more on the most popular exhibits, see David E. Nye, "Ritual Tomorrows: The New York World's Fair of 1939," *History and Anthropology*, 6:1, 1992, pp. 1–21.

Chapter 8

1 In March, 1941 a Gallup Poll found that an overwhelming 83 percent of the American people opposed entering World War II. *Gallup Poll: Public Opinion, 1935–1971*, Vol. 1 (New York: Random House, 1972), p. 271.

2 Market Analysts, Inc. interviewed 1,020 people in August, 1939 and presented the results to fair managers in, "Attendance and Amusement Area Survey of the New York World's Fair 1939, Incorporated," New York Public Library, World's Fair Collection, 1934–40, p. 11.

3 For a survey of recent work, see John E. Findling, ed., *Historical Dictionary of World's Fairs and Expositions, 1851–1988* (Westport: Greenwood Press, 1990). For a comparative overview of French, English and American fairs, see Paul Greenhalgh, *Ephemeral Vistas* (Manchester: University of Manchester Press, 1988). On American fairs see Robert Rydell, *All the World's a Fair* (Chicago: University of Chicago Press, 1984), and *World of Fairs* (Chicago: Chicago University Press, 1993). For the beginnings of a theory to deal with fairs, see Burton Benedict, *The Anthropology of World's Fairs: San Francisco's Panama Pacific International Exposition* (London: Scholar Press, 1983).

4 Spectacular lighting experts moved from one fair to the next, for example, and often the same architects participated in one exposition after another.

5 The Crystal Palace was also heir to a previous fair tradition, as Greenhalgh notes.

6 See Francis V. O'Connor, "The Usable Future," in Helen Harrison, ed., *Dawn of a New Day: The New York World's Fair, 1939/40* (New York: Queen's Museum and New York University Press, 1980).

7 The New York fair's "Temple of Religion" was hidden on all sides behind much larger pavilions, including "Gas Exhibits" and "Beech-Nut." In the many articles about the fair, the religion building is virtually never mentioned, suggesting again the secular nature of the event. The building was primarily used for concerts and lectures.

8　Reference to the war could not be avoided in the second year of the fair, but even then it was downplayed.

9　See Folke Kihlstedt, "Utopia Realized: The World's Fairs of the 1930s," in Joseph Corn, ed., *Imagining Tomorrow: History, Technology, and the American Future* (Cambridge: MIT Press, 1986).

10　See Rydell, pp. 49–68.

11　Robert Rydell, "The San Francisco Golden Gate International Exposition and the Empire of the West," in Rob Kroes, ed., *The American West as seen by Europeans and Americans* (Amsterdam: Free University Press, 1989), pp. 342–60.

12　See Greenhalgh, pp. 27–41.

13　At San Francisco nations from the Pacific rim appeared, while the only nations from the Far East in New York were Japan and Australia, neither of which made a major investment in their exhibits.

14　See Findling, p. 293. The theme of the fair was further defined as the result of a large dinner at the New York Civic Club in 1935, which formed a committee to develop the idea.

15　The American past appeared as a theme in the pavilions of the States more than anywhere else, and this part of the fair appears to have been the least visited.

16　One immediate inspiration for the New York fair was the financial and popular success of the Chicago Century of Progress Exposition of 1933–34. *Century of Progress, Official Guide*, second edition, 1933. The illumination was done by W. D'Arcy Ryan, the same man responsible for illuminating Niagara Falls. See W. D'Arcy Ryan, "Illumination of A Century of Progress Exposition, Chicago, 1933," *General Electric Review*, May 1934, pp. 227–231.

17　Larry Zim, Mel Lerner and Herbert Rolfes, *The World of Tomorrow, 1939, New York World's Fair* (New York: Harper and Row, 1988) reprints several interesting contemporary accounts, but has little text of its own. Warren Susman later revised an essay from the Harrison volume as chapter 11 of *Culture as History* (New York: Pantheon, 1985). Another important source is Stanley Appelbaum's selection of photographs of the fair: *The New York World's Fair* (New York: Dover, 1979). Two other important essays are: Folke Kihlstedt's "Utopia Realized: The World's Fairs of the 1930s," in Joseph Corn, ed., *Imagining Tomorrow* (Cambridge: MIT Press, 1986), and the last chapter of Alice Goldfarb Marquis, *Hopes and Ashes: The Birth of Modern Times*, (New York: Free Press 1986).

18　Market Analysts, Inc., pp. 4–8, 10.

19　The following had their own pavilions: Argentina, Brazil, Canada, Chile, France, Finland, Great Britain, Italy, Japan, The Netherlands, Norway, Portugal, Poland, Romania, Switzerland, Sweden, Turkey, USSR, Czechoslovakia, and Venezuela. The Czech pavilion remained open despite the fact that Germany had swallowed up the country before the fair began. See "Czech Exhibit is Speeded," *New York Times*, 16 May 1939.

20　From Grover Whalen, *Mr. New York* (New York: Putnam, 1955). Cited in Findling, p. 297.

21 See Ivan Karp and Steven D. Lavine, *Exhibiting Cultures: The Poetics and Politics of Museum Display* (Washington: Smithsonian Institution Press, 1991), particularly pp. 11–21 and Kenneth Hudson, "How Misleading Does an Ethnographical Museum Have to Be?" pp. 457–65,

22 *New York World's Fair: Official Guide* (New York: Exposition Publications, 1939). In practice, Japan was obviously given a waiver of this rule in its exhibit, but most other nation's complied.

23 "The New York Fair," *Architectural Forum*, June, 1939, p. 395.

24 "Italy's Huge Unit is 100 percent Complete," *New York Times*, 18 May 1939. On Mussolini's proposal to duplicate Coney Island, see *New York Times*, 16 July 1939, p. 21. *Architectural Forum* (June, 1939), p. 456.

25 "French Pavilion, Unready, Popular," *New York Times*, 16 May 1939. The British building was almost as popular as the Russian. The French exhibit was not yet fully open at this time so it cannot be compared. *New York Times*, 17 May 1939.

26 *Architectural Forum*, June 1939, p. 448.

27 "France's Impressive Pavilion is Dedicated," *New York Times*, 25 May 1939, p. 22. *Architectural Forum*, June 1939, p. 453.

28 It was loaned by Lincoln Cathedral and when it arrived was transported behind bullet-proof glass by a motorcade of New York police from the harbor to the fairground. *New York Times*, 22 April 1939, p. 3.

29 For Washington memorials, see *Times, Special Supplement on the World's Fair*, 30 April 1939, p. 29. "Oak from Britain Is Planted at Fair," *New York Times*, 19 May 1939. Speeches given at the event were explicit in calling the tree a symbol of Anglo-American friendship and shared values of "order, freedom and justice in a world shaken by upheaval."

30 In *New York Times*, 23 April 1939.

31 On the Japanese exhibit, see *New York World's Fair, 1939 Official Guide* (New York: Exposition Press, 1939), pp. 107–108.

32 "Fair Almanac: The Main Events" Special Supplement, *New York Times*, 30 April 1939. *New York Times*, 2 June 1939. France's day was Bastille Day, The Netherlands celebrated their Queen's birthday at the fair, and so forth. For some reason, Poland was awarded a whole week at the fair in October, by which time the country had already been absorbed by Germany.

33 "Irish Exhibit Spins Romance of Linen," *New York Times*, 17 May 1939. The South Africans did not have a national pavilion, but instead put millions of dollars worth of diamonds into the "House of Jewels," which was at the center of the fairground.

34 *New York Times*, 11 June 1939.

35 See Greenhalgh for analysis of the convoluted reasoning used to explain how, taken together, the Magna Carta, the creation of an empire, and the American Revolution posed no inconsistencies, pp. 135–6.

36 Greenhalgh, p. 137.

37 "French Art Works Reflect 500 Years," *New York Times*, 25 May 1939.
38 "Huge Steel Statue Tops Soviet Center," *New York Times*, 17 May 1939, p. 20. Greenhalgh notes, "A model of the Palace of the Soviets in semi-precious stones attracted a lot of attention", p. 136.
39 *Architectural Forum*, June 1939, p. 459.
40 Dutch pavilion, see *New York World's Fair: Official Guide*, p. 110.
41 World's Fair supplement, *New York Times*, 30 April 1939.
42 "Huge Steel Statue Tops Soviet Center," *New York Times*, 17 May 1939, p. 20.
43 "Thinking Watch Is Shown at Fair," *New York Times*, 19 May 1939, p. 18. Also see Zim, p. 163.
44 "France's Impressive Pavilion is Dedicated," *New York Times*, 25 May 1939, p. 22.
45 This is the story of a non-commissioned officer of the Hungarian Hussars who so distinguished himself in the Napoleonic wars that he became a general, was admitted to court life in Vienna, and was engaged to marry the daughter of the Austrian emperor, only to reject the offer and return to the simple life of his village, where he married his boyhood sweetheart. Its premier in Budapest took place in 1926, but in 1939 it had never been performed in the US.
46 No one was killed. "Pilot Boat Sunk by Prince's Ship," *New York Times*, 28 April 1939.
47 Benedict, *The Anthropology of World's Fairs*.
48 *New York Times*, 13 Jan. 1939, p. 1; 2 Feb. 1939, p. 5.
49 David E. Nye, "Yesterday's Ritual Tomorrow: The New York World's Fair of 1939," in *Anthropology and History*, 6:1, 1992, pp. 1–21.
50 Roland Marchand, "Corporate Imagery and Popular Education: World's Fairs and Expositions in the United States, 1893–1940," in David E. Nye and Carl Pedersen, eds. *Consumption and American Culture* (Amsterdam: Free University Press, 1991), pp. 21–3.

Chapter 9

1 Vannevar Bush, *Science—The Endless Frontier* (Washington: National Science Foundation, 1945). Walter A. McDougall, *The Heavens and the Earth: A Political History of the Space Age*. (New York: Basic Books, 1985). John F. Kennedy, *"Let the Word Go Forth": The Speeches, Statements and Writings of John F. Kennedy* ed. Theodore C. Sorensen (New York: Delacorte Press, 1988), pp. 100–102.
2 See Rip Bulkeley, *The Sputnik Crisis and Early United States Space Policy* (London: Macmillan, 1991).
3 This analogy was used by Wernher von Braun and has often been repeated. For autobiographies, see, for example, Gene Farmer and Dora Jane Hamblin, *First on the Moon, A Voyage With Neil Armstrong, Michael Collins, Edwin E. Aldrin, Jr.* (London: Michael Joseph, 1970).

4   William Irwin Thompson, *Passages About Earth: An Exploration of the New Planetary Culture* (New York: Harper and Row, 1973), pp. 1–9.

5   I will be using that term roughly in the sense advanced by Daniel Dayan and Elihu Katz, in *Media Events: The Live Broadcasting of History* (Cambridge: Harvard University Press, 1992).

6   Tom Wolfe, *The Right Stuff* (New York: Farrar, Straus and Giroux, 1979).

7   See, for example, Alan Shepard and Deke Slayton, *Moon Shot: The Inside Story of America's Race to the Moon* (Atlanta: Turner Publishing, 1994).

8   Fuller & Smith & Ross, Inc., "Attitudes Toward the Moon Race Among Opinion Leaders and The General Public," Section IV, p. 1. Copy in the National Aeronautics and Space Administration Library, Washington DC.

9   The Department of Defense was against a separate space agency from the beginning and did its best to colonize it afterwards.

10 Fuller & Smith & Ross, IV, p. 2.

11 2 April 1948, *Gallup Poll: Public Opinion, 1935–1971*, Vol. 1 (New York: Random House, 1972), pp. 27, 722.

12 Harris Poll in *Washington Post*, Monday, 1 Nov. 1965.

13 Harris Poll in *Washington Post*, Monday, 31 July 1967.

14 Harris Poll in *Philadelphia Inquirer*, 17 Feb. 1969. Opinions were more positive in response to the question: "Do You Favor or Oppose the Moon Landing?"

	for	oppose	don't know
By education			
8th grade	19	65	16
high school	37	51	12
college	62	28	10
By gender			
men	46	43	11
women	32	54	14
By age			
under 35	51	39	10
35–49	43	46	11
50 and over	28	57	15

15 *Washington Post*, Monday, 14 July 1969.

16 *Washington Post*, 25 August 1969.

17 "49 percent Oppose Moon Project," *Philadelphia Inquirer*, 17 Feb. 1969.

18 Interview with Mrs. Joyce Rogers, *San Jose Mercury*, 7 March 1969. Several others of the seven interviewed expressed similar sentiments about poverty programs. Only two men were entirely supportive.

19 Michael Smith, "Selling the Moon: The U.S. Manned Space Program and the Triumph of Commodity Scientism," in T. J. Jackson Lears and Richard Wightman Fox, eds., *The Culture of Consumption* (New York: Pantheon, 1983), pp. 175–209.

20 *Los Angeles Herald–Examiner*, 20 July 1969.

[21] *The Philadelphia Inquirer,* 17 Feb. 1969.

[22] *The Chicago Daily Defender,* 12 July 1969. *Muhammad Speaks,* 25 July 1969.

[23] *Amsterdam News,* 12 July 1969.

[24] Harris Poll in *Philadelphia Inquirer,* 17 Feb. 1969.

[25] "Man on the Moon," *Amsterdam News,* 12 July 1969, pp. 1, 48.

[26] See *Baltimore Afro-American,* 19 July 1969, p. 1; *Chicago Daily Defender,* 12 July 1986, p. 6, editorial page. *Muhammad Speaks,* 25 July 1986, p. 3.

[27] *New York Times,* 16 July 1969, p. 22; *Washington Post,* 7 July 1969.

[28] *Washington Post,* 21 July 1969.

[29] Drew Pearson, *Washington Post,* 10 July 1969, p. F11.

[30] "Americans Still Question Space Budget," *Washington Post,* 25 August 1969.

[31] *USA Today* poll, taken 14 July 1969, cited in Ordway and Liebermann, *Blueprint for Space.* The Harris Poll reached the same conclusion a few weeks later, *Washington Post,* 25 August 1969.

[32] On the crowd's experience of space launches, see David E. Nye, *American Technological Sublime* (Cambridge: MIT Press, 1994), pp. 239–241, 243–251. For the flavor of popular enthusiasm, see Brian Horrigan, "Popular Culture and Visions of the Future in Space, 1901–2001," in Bruce Sinclair, ed., *New Perspectives on Technology and American Culture* (Philadelphia: American Philosophical Society, 1986).

[33] All early broadcasts were in black and white, because color equipment added too much weight. The Mercury flights carried only photographic equipment, while television appeared in the Gemini series. John Glenn's orbiting mission carried a film camera whose footage was later broadcast. Not until Apollo X were spacecraft fitted out with color cameras, but because of weight considerations, black and white had to suffice for the landing module. Courtney G. Brooks, et al., *Chariots for Apollo: A History of Manned Lunar Space Craft* (Washington: NASA History Series, 1979), p. 329.

[34] Dayan & Katz, pp. 6–9.

[35] Harris Poll in *Philadelphia Inquirer,* 17 Feb. 1969.

[36] Dayan & Katz, p. 126.

[37] "It was History—but it was also a Great Day for the Beach," *Honolulu Advertiser,* Red Stripe Sunrise Edition, 21 July 1969, Sec A–7, p. 1. This example courtesy of Robert Baehr.

[38] My colleague, Joyce Pedersen, was a graduate student in history at Berkeley. I was at the University of Minnesota.

[39] Dale Carter, *The Final Frontier* (London: Verso, 1988), pp. 198–199 and passim.

[40] On youth reaction, see "Teenagers Show Cynicism on America's Space Race," *New Orleans Times–Picayune,* 8 Feb. 1968.

[41] *Washington Star,* 13 July 1969, D2. Of course the term "square" was current in the 1950s and was already a bit old-fashioned by 1969.

[42] Nevertheless, it should be noted that Nixon did not pour money into NASA. A mission to Mars would take far too long to do him any good politically, and as

Dale Carter pointed out to me, Nixon felt that outer space was strongly associated with Kennedy.

43 Dayan & Katz, p. 37.

44 Ibid., p. 44.

45 If the actual moon landing by itself might be considered a "conquest," Dayan and Katz argue that the space flights of the 1960s collectively illustrate three forms of the media event. "They began as a Contest between the United States and the Soviet Union when the first Sputnik was launched into space. Ten years later, as advertised, came the American Conquest of the moon. Finally (but long anticipated) the astronaut heroes were crowned and recrowned by society and the media." Dayan and Katz, p. 27. [Capitalization in original.]

46 *Washington Post*, 22 July 1969.

47 Norman Mailer, who has some engineering education, understood, if he did not condone, this instrumental use of language. For NASA only the machine itself was sublime, not what was said about it. Norman Mailer, *Of a Fire on the Moon* (London: Weidenfield and Nicolson, 1970).

48 Robert Baehr, "The Moon: How Far Is It. . . to Charles Lindberg?" in David E. Nye and Christen Kold Thomsen, *American Studies in Transition* (Odense: Odense University Press, 1985), pp. 153–174.

49 J. B. Jackson, *Discovering the Vernacular Landscape* (New Haven: Yale University Press, 1984), p. 8.

50 Wyn Wachhorst, "Seeking the Center at the Edge: Perspectives on the Meaning of Man in Space." *The Virginia Quarterly Review* 69:1, Winter, 1993, pp. 8–9.

51 Patricia Nelson Limerick, "The Adventures of the Frontier in the Twentieth Century," in James R. Grossman, ed., *The Frontier in American Culture* (Berkeley: University of California Press, 1994), p. 88.

Chapter 10

1 Quotation from Jean-Francois Lyotard, "Rules and Paradoxes and Svelte Appendix," in *Cultural Critique* 5, Winter, 1986–1987, p. 210. On postmodernism, see Lyotard's *The Postmodern Condition: A Report on Knowledge* (Minneapolis: University of Minnesota Press, 1984, translation of *La Condition Postmoderne* [Paris: Les Editions de Minuit, 1979]). Ihab Hassan, "The Culture of Postmodernism," *Theory, Culture, Society*, 1985, 2(3), pp. 123–124. David Harvey, *The Condition of Postmodernity* (Oxford: Basil Blackwell, 1989). On the history of computing until 1970, see Michael R. Williams, *A History of Computing Technology* (Englewood Cliffs: Prentice-Hall, 1985).

2 For a critique of Baudrillard that shows how typical he is of a long line of European intellectuals, see Rob Kroes, "Flatness and Depth: Two Dimensions of the Critique of American Culture in Europe," pp. 228–239, in David E. Nye

and Carl Pedersen, eds. *Consumption and American Culture* (Amsterdam: University of Amsterdam Press, 1991).

3   Kurt Vonnegut, *Player Piano* (New York: Avon Books, 1952). This view of the effects of technology's future consequences was a common theme in earlier dystopian science fiction. See, for example, H. G. Wells, *When the Sleeper Wakes*, reprinted in *Three Prophetic Novels* (New York: Dover, 1960).

4   Cited in Walter M. Mathews, ed., *Monster or Messiah? The Computer's Impact on Society* (Jackson: University of Mississippi Press, 1980), p. 22.

5   Richard S. Tedlow, *New and Improved: The Story of Mass Marketing in America* (New York: Basic Books, 1990), p. 348.

6   Later, in the third period the computer will be as ubiquitous as the telephone is today. Then it no longer seems an intrusion but a "natural" and unremarkable part of the social background, as electric lights were by the 1950s. Computers will be noticed only when they break down.

7   Lyotard, *The Postmodern Condition*, p. 4.

8   Cited in Tedlow, p. 348.

9   Women were quite commonly employed on assembly lines. See Nye, *Electrifying America*, chapter 5.

10 For an authoritative analysis of this process by a former worker turned university professor, see Harley Shaiken, *Work Transformed: Automation and Labor in the Computer Age* (Lexington: Lexington Books, 1986).

11 This section of my argument relies on Shoshana Zuboff, *In the Age of the Smart Machine: The Future of Work and Power* (New York: Basic Books, 1988).

12 Ibid., p. 344.

13 Ibid., p. 348.

14 Ibid., p. 351.

15 On the problems of assembly-line mass production and the development of high-tech cottage industry suited to small batch production, see Charles F. Sabel, *Work and Politics: The Division of Labor in Industry* (Cambridge: Cambridge University Press, 1982), chapter 5.

16 Zuboff, p. 356.

17 Ibid., p. 359.

18 Michael Shallis, *The Silicon Idol: The Micro Revolution and its Social Consequences* (Oxford: Oxford University Press, 1985), p. 115.

19 Hassan, pp. 123–124.

20 Diane Ackerman, *The Natural History of the Senses* (New York: Random House, 1990), p. 165.

21 All reported in *Facts on File*, 1988.

22 A teller did precisely this, see Stanley Rothman and Charles Mosmann, *Computers and Society*. 2nd ed., (Palo Alto: Science Research Associates, 1976), p. 353.

23 Harvey, pp. 48–49.

Conclusion

1   Albert Boime, *The Magisterial Gaze: Manifest Destiny and American Landscape Painting*, c. *1830–1865*, Washington: Smithsonian Institution, 1991, p. 75.

2   Pierre Macherey, *A Theory of Literary Production* (London: Routledge & Kegan Paul, 1978), p. 155.

3   William Bradford, *Of Plymouth Plantation*, ed. Samuel Eliot Morison, (New York: Knopf, 1952), chapter 9.

4   Such contradictions between symbolic action and behavior map a fundamental tension within American culture. This tension between the land as idealized, abstract space and its restructuration through technology has long been of central concern to scholars. Most notably, two classic studies, Henry Nash Smith's *Virgin Land* (Cambridge: Harvard University Press, 1950) and Leo Marx's *The Machine in the Garden* (New York: Oxford University Press, 1964) have engaged this problem.

5   Michael L. Smith, "Recourse of Empire: Landscapes of Progress in Technological America," in Merritt Roe Smith and Leo Marx, *Does Technology Drive History? The Dilemma of Technological Determinism* (Cambridge: MIT Press, 1994), pp. 40–42.

6   Marshall Everett, *The Book of the Fair: The Greatest Exposition the World has Ever Seen, A Panorama of the St. Louis Exposition* (St Louis: Henry Neil, 1904), p. 169.

7   Ibid., pp. 172–174.

8   See David E. Nye, *Electrifying America: Social Meanings of a New Technology* (Cambridge: MIT Press, 1990), chapter 8; and Roland Marchand, "Corporate Imagery and Popular Education: World's Fairs and Expositions in the United States, 1893–1940," in David E. Nye and Carl Pedersen, eds. *Consumption and American Culture* (Amsterdam: Amsterdam Free University Press, 1991), pp. 18–33.

9   See John M. Findlay, *Magic Lands: Western Cityscapes and American Culture After 1940* (Berkely: University of California Press, 1992).

10   For a discussion of the myth of the eternal return, see Mircea Eliade, *The Sacred and the Profane* (New York: Harcourt, Brace, and World, 1959), pp. 74–80, *passim*. For application to national parks, see John Sears, *Sacred Places: American Tourist Attractions in the Nineteenth Century* (New York: Oxford University Press, 1989, pp. 12–13, 38–39.

11   United States Department of the Interior, *The Colorado River: A Comprehensive Report on the Development of the Water Resources of the Colorado River Basin for Irrigation, Power Production, and Other Beneficial Uses in Arizona, California, Colorado, Nevada, New Mexico, Utah, and Wyoming*. Washington, March 1946, p. 25.

12   Jacques Ellul, *The Technological Society* (New York: Alfred A. Knopf, 1964).

13   Michel Foucault, *Madness and Civilization* (Random House, 1965); Michel Foucault, *Birth of the Clinic* (Pantheon, 1973); Michel Foucault, *Power/Knowledge: Selected Interviews and Other Writings*, ed. Colin Gordon (Pantheon, 1980), pp. 153–156.

[14] Mike Davis, *City of Quartz* (New York: Vintage, 1992), p. 227.

[15] J. B. Jackson, "The Abstract World of the Hot-Rodder" in Ervin H. and Margaret J. Zube, eds., *Changing Rural Landscapes* (Amherst: University of Massachusetts Press, 1977), pp. 146–147.

[16] See Steven Kern, *The Culture of Space and Time, 1880–1918* (Cambridge: Harvard University Press, 1983), pp. 31–39, *passim*.

[17] Marshall McLuhan, *Understanding Media: The Extensions of Man* (New York, McGraw-Hill, 1964)

[18] Joshua Meyrowitz, *No Sense of Place: The Impact of Electronic Media on Social Behaviour* (Oxford University Press, 1985), p. 309.

[19] Nicholas Negroponte, *Being Digital* (New York: Vintage, 1995), p. 229.

[20] Newt Gingrich, *To Renew America* (New York: HarperCollins, 1995); Alvin and Heidi Toffler, *Creating a New Civilization: The Politics of the Third Wave* (Atlanta: Turner Publishing, 1994).

[21] Hayden White, *Metahistory: The Historical Imagination in Nineteenth Century Europe* (Baltimore: Johns Hopkins University Press, 1973), p. 10.

INDEX